Editing the Soul

AnthropoScene
THE SLSA BOOK SERIES

Lucinda Cole and Robert Markley, General Editors

Advisory Board:
Stacy Alaimo (University of Texas at Arlington)
Ron Broglio (Arizona State University)
Carol Colatrella (Georgia Institute of Technology)
Heidi Hutner (Stony Brook University)
Stephanie LeMenager (University of Oregon)
Christopher Morris (University of Texas at Arlington)
Laura Otis (Emory University)
Will Potter (Washington, D.C.)
Ronald Schleifer (University of Oklahoma)
Susan Squier (Pennsylvania State University)
Rajani Sudan (Southern Methodist University)
Kari Weil (Wesleyan University)

Published in collaboration with the Society for Literature, Science, and the Arts, AnthropoScene presents books that examine relationships and points of intersection among the natural, biological, and applied sciences and the literary, visual, and performing arts. Books in the series promote new kinds of cross-disciplinary thinking arising from the idea that humans are changing the planet and its environments in radical and irreversible ways.

Editing the Soul

Science and Fiction in the Genome Age

Everett Hamner

The Pennsylvania
State University Press
University Park,
Pennsylvania

"Gay Gene Isolated, Ostracized" in chapter 1 originally appeared in the "News in Brief" section of *The Onion* 44.26. 9 April 1997.
Reprinted with permission of *The Onion*.
Copyright © 2017, by Onion Inc.

Library of Congress Cataloging-in-Publication Data

Names: Hamner, Everett, 1975– , author.
Title: Editing the soul : science and fiction in the genome age / Everett Hamner.
Description: University Park, Pennsylvania : The Pennsylvania State University Press, [2017] | Series: AnthropoScene: the SLSA book series | Includes bibliographical references and index.
Summary: "An interdisciplinary exploration of how genetic engineering is transforming our narratives about the core of human personhood, and how those narratives are shaping official policies"—Provided by publisher.
Identifiers: LCCN 2017033621| ISBN 9780271079325 (cloth : alk. paper) | ISBN 9780271079332 (pbk. : alk. paper)
Subjects: LCSH: Genetic engineering in literature. | Cloning in literature. | Science fiction—History and criticism.
Classification: LCC PN3433.6 .H364 2017 | DDC 809.3/8762—dc23
LC record available at https://lccn.loc.gov/2017033621

Copyright © 2017 Everett Hamner
All rights reserved
Printed in the United States of America
Published by
The Pennsylvania State University Press,
University Park, PA 16802-1003

The Pennsylvania State University Press is a member of the Association of American University Presses.

It is the policy of The Pennsylvania State University Press to use acid-free paper. Publications on uncoated stock satisfy the minimum requirements of American National Standard for Information Sciences—Permanence of Paper for Printed Library Material, ANSI Z39.48-1992.

The interesting thing about hope is that if you don't have any hope then there's less hope. It is a self-generating thing. So if you do have it, there's likely to be more.

—Margaret Atwood

Say it has happened already, just the way it will.

—Richard Powers

Contents

Acknowledgments | ix

Introduction: *Regenesis* | 1

1. Genetics as Science, Ideology, and Fiction | 31

2. The Evolution of Genetic Fantasy | 59

3. The Cultural Determinism of Genetic Realism | 91

4. Serpent Women, Prophets, and Satire in Genetic Metafiction | 133

5. The Predisposed Agency of Genetics and Fiction | 175

Coda: *Arrival* | 209

Notes | 225

Works Cited | 237

Index | 249

Acknowledgments

I would never have been in a position to begin this project without the investments of numerous academic mentors. I doubt Yucaipa High School (CA) has ever offered its students a stronger combination of English teachers than the eyebrow-raising David Braxton, the inestimable Scott Smith, and the multitalented Tom Roper, who also taught me AP Physics. Neil Hertz and Robert Massa believed in me during and after my undergraduate years at Johns Hopkins, and in my first years of teaching at the Boys' Latin School of Maryland, Kay Schuyler pushed me to new heights. At Regent College in Vancouver, Maxine Hancock, Charles Ringma, John Stackhouse, the late Roger Lundin, and especially Loren and Mary Ruth Wilkinson—godparents to my children—were immeasurably influential. In the University of Iowa's stellar English department, my many hours with dissertation committee members Lori Branch, David Klemm, Brooks Landon, and Harry Stecopoulos and unparalleled codirectors Claire Fox and Garrett Stewart were wonderfully transformative. During these years, I was immensely grateful for support from the Hodson Trust, the Woodrow Wilson Foundation, and the University of Iowa Graduate College.

My interdisciplinary habits have grown significantly through the opportunity to teach widely across literary, film, science, and religious studies at Western Illinois University (WIU). This book was composed in large measure during a sabbatical, the release for which I am indebted to my chair, Mark Mossman, and dean, Susan Martinelli-Fernandez, as well as the provost, president, and board of trustees. Among many other dedicated colleagues at WIU, I am especially thankful for support from Kristi Mindrup, Emily Pitz, Joe Rives, Alison Shook, Jim Schmidt, and Tom Finley and long-running conversations with David Banash, Amy Carr, Andrea Hyde, Amy Mossman, Jim Patterson, Jim Rabchuk, Shazia Rahman, Susan Romano, and Tammy Werner. *Editing the Soul* is also hard to imagine without having been able to teach several courses on its subject matter. A huge thanks to all the students who made those classes so rich, especially Allyson Borkgren, Joseph Boyle, Chelsea Clearman, Kirsten Dillender, Aaron Graf, Sarah Horowitz, Matthew Lewerenz, Raymond Norris, Chris Sanders-Ring, Kris Souza, and Gina Wilkerson. I also want to honor contributions in other courses from Zach Almquist, Cait Bodenbender, Hayleigh Covella, Bryan Dietsch, Steve Long, Julia McMeekan, and Nikki Steinbaugh.

Academia thrives when leading scholars choose to invest in junior faculty well beyond their own institutions. My list of mentor friends begins with Tracy Fessenden, Clark Gilpin, Philip Goff, Paul Harvey, and everyone at the Young Scholars in American Religion program at Indiana University–Purdue University Indianapolis, including the members of my 2009–11 cohort. Ronald M. Green has long provided significant encouragement for this project, and more recently, Regenia Gagnier and John Dupré opened doors for me to expand my relationships with UK scholars. I have also received valuable encouragement and advice over the years from Michael Bérubé, Jay Clayton, Istvan Csiscery-Ronay Jr., Paul Gutjahr, Michael Kaufmann, Susan Squier, and Sherryl Vint. Finally, Priscilla Wald has been one of my most influential supporters and interlocutors for a decade now. I cannot imagine this book existing without her scholarship or her personal commitment as an informal adviser and friend.

There are far too many colleagues who have shaped my thinking and fueled my efforts for me to list here. Still, among my friends in literary studies, I must mention Angela Balla, Dan Boscaljon, Gerry Canavan, Mike Chasar, Liam Corley, Ann Jurecic, Heidi Kim, Kevin Seidel, and Chad Wriglesworth. Rebekah Sheldon, you get an extraspecial shout-out for hard root beer ice cream floats and all they symbolize. Among the many friends in religious studies and history whose insights have strengthened my work, Ed Blum, Nathaniel Comfort, Charles Irons, Lynn Neal, Jon Ebel, Maura Farrelly, Nancy Menning, and Matthew Sutton stand out. Finally, among the scientists who have encouraged my scholarship, I want to thank especially Salman Hameed, Chuck Lydeard, and fellow Darwin Dayers John Logsdon and Maurine Neiman.

One element of academia I often lament is real and perceived divisions between "creative" and "critical" writers. Contra that narrative, I have been enormously thankful to have broken bread with several authors and other artists whose work features heavily in this book, including Margaret Atwood, Jeffrey Eugenides, Cosima Herter, Ursula K. Le Guin, Graeme Manson, and

Richard Powers. For anyone who's curious, these people are all just as gracious and brilliant as they appear from a distance.

I often utilize auteur theory in studying film, but just as a director is far from a film's only creator, I am hardly the only wordsmith with fingerprints on these pages. While responsibility for errors is entirely my own, Andrew Hoogheem and Dan Malachuk were especially crucial dialogue partners, early readers, and dog-walking companions. To Dan, I must add this pithy note: I could not ask for a better friend, teacher, and scholar with whom to partner at WIU-Quad Cities.

I selected Penn State University Press (PSUP) for this project because of my trust in longtime Society for Science, Literature, and the Arts (SLSA) leaders and AnthropoScene series coeditors Lucinda Cole and Robert Markley and my immediate affinity for its absolutely stellar editor in chief, Kendra Boileau. Along the way, I have also much appreciated the contributions made by PSUP's two anonymous readers, as well as Brendan Coyne, Jennifer Norton, Laura Reed-Morrison, Alex Vose, and Megan Grande. If you are a prospective author, PSUP gets my heartiest endorsement.

I have never been a brain on stilts, so there are others to thank whose impact was less direct but no less crucial. My life on the baseball field has long fed and provided release from my intellectual life, so I want to acknowledge here all the families of the 2013–16 Barnstormer Bolts and the 2017 Diamond Kings. At an even deeper level—and as absurd as it feels to try to sum up their significance so briefly—my parents, Glen and Cheryl Hamner, and my sister, A. J. Hamner, have shaped every word on these pages. As a child, I was fortunate to have a mom who viewed me with fascination, a dad who had no greater priority, and a sister who cheered for me with greater intensity than anyone. As a reader, a viewer, and a writer in these pages, I speak for myself, but always with an ear developed during our hours reading Lloyd Alexander aloud and arguing about the implications of *Star Trek* (mostly *The Next Generation*).

This book goes to press as the twentieth anniversary of my marriage with Jennifer Lynn Hamner (née Pummell) draws near. (In 1997, we were already feminist enough to think about hyphenating last names, but the result struck us as too violent.) Nonetheless, I am grateful to the Pummells for the trust they've given me and for all the times they've understood when I had to work. Jenn, you know better than anyone that this book is the product of your endurance as much as mine. It is one result of your readiness to take a long series of nearly absurd risks toward the possibility of my gainful academic employment, and it would not exist without your repeated insistence on enabling conference papers, chats with colleagues, research trips, and hours hidden away to write. Gratitude doesn't begin to encompass the feeling.

To Aidan and Rowan: you have also formed these pages. You are wonderfully imperfect, impossibly beautiful, startlingly insightful gifts who make me want to be my best. There is so much that is biologically material between us, but it is barely describable in those terms. I could never have imagined that I would once laugh at a 3 AM diaper change turned scatological art exhibition or, much more often (thankfully), that I would find so much joy in being still and reading with you. The nurture between us has always been two way, and I am immensely proud of your ever-growing (if always predisposed) agency.

Introduction
Regenesis

In 2012, a guest on *The Colbert Report* brought one of the more unique "show and tell" items in late-night television history. Writers making the talk-show rounds routinely send advance editions of new books, but geneticist George Church upped the ante. On a tiny piece of paper, in a single drop of liquid, he handed the comedian twenty million copies of his book, entirely encoded in the A, G, C, and T base-pair language of DNA. As *Time* reported later, Church actually "printed" many more copies at his lab—some seventy billion more. Gifted with an unusual combination of scientific intuition, entrepreneurial instinct, and winsome showmanship, he nonchalantly explained, "We calculated the total copies of the top two hundred books of all time, including *A Tale of Two Cities* and the Bible and so on, and they add up to about twenty billion. We figured we needed to go well beyond that."

These are stunning numbers, as is the more basic fact that humanity is learning not only to read the language of DNA but also to write with it. Even many scientists are surprised at the pace at which genetics has accelerated. Enthusiastic about the "room at the bottom," physicist Richard Feynman challenged readers in 1959 to render a page of text at 1/25,000 scale. Half a century later, Church reproduced a fifty-three-thousand-word book and its eleven images on a dot smaller than Stephen Colbert's fingernail, working at roughly a 1/10,000,000 scale. Consider the storage capacity of your phone or computer (a few dozen or a few hundred gigabytes in most cases) and then imagine a geneticist installing 455 *billion* gigabytes in one gram of solution. Reading and writing at the genetic level is redefining the word *small*—and while awe-inspiring, such realizations can also be profoundly discomfiting.

Much to his audience's delight, Colbert's hyperconservative character immediately gestured to that anxiety, straight-facedly inquiring of Church, "How do you think your work . . . will eventually destroy all of mankind?"

After showing off a traditional hardbound version of Church's book, *Regenesis: How Synthetic Biology Will Reinvent Nature and Ourselves*, Colbert went on, "Do we need reinventing? We were invented once. By God, the Almighty Father, Maker of Heaven and Earth. Are you playing God, sir? 'Cause you certainly have the beard for it." Beneath their levity, Colbert's questions reflected deep concerns felt by many observers of contemporary biotechnology, including very well-informed ones. The ability to compress untold amounts of data into a drop of solution might not seem especially threatening, but once the potential applications of new genetic technologies expand to projecting individuals' likelihood of manifesting life-threatening diseases, selecting attributes for one's children, reviving extinct species like the woolly mammoth, or combining traits from existing species to create "transgenic" organisms, the worries begin to mount.

It should not be surprising that Colbert, who is in reality liberally Catholic, introduced theological language to the conversation, and only partly in jest. To discuss possibilities afforded by stem cell research, genome testing, and gene editing is to invite questions about personhood, spirituality, and the very essence of individual identity. Thus *Editing the Soul*'s overarching questions: How is contemporary literature and film about genetics—from the most fantastic to the most carefully researched—reimagining the nature of our souls, the influences that shape them, and the possible relationships between self and other? Conversely, how are new biological discoveries and tools changing the very structures and purposes of fiction, whether in literary, cinematic, or other form? These are large questions that can lead to very abstract, metaphysical answers. It is understandable that some respond, "How about we just stick to the data, make people as healthy as we can, and leave the debates to the philosophers and theologians?" The problem is that not even scientists can remain entirely empirical, as much as their laboratory labors require the effort; discussing possibilities for genetic testing and modification requires the infinite variables of language, with all its porous boundaries between the physical and the spiritual, the scientific and the mystical.

To explore and especially to remake the nature of our species is inevitably to invite revisions of our foundational mythologies, the stories by which our species lives—consciously and otherwise. It is perhaps inevitable, then, that those who are most optimistic about genetic testing and pharmacogenomics improving medicine would draw on metaphors and narratives that promise individual transcendence, consciously and otherwise. Of course, sometimes the overstatements can be chalked up to the bite-sized summaries by which biotechnological possibilities are simplified for public consumption. Setting aside the problem of media misrepresentations, though, scientists themselves are under increasing pressure to sensationalize their work's potential impact. The heightened metaphysical rhetoric often surrounding new genetic discoveries and biomedical applications is not just a product of format but also a reflection of audience expectations. Because scientists rely on cultural narratives to describe their work's significance, it should not be surprising that fiction about genetics would both reflect and influence the goals and self-understandings of actual scientists. It is not that genetic fiction decides what scientists do in the lab on a day-to-day basis; rather, when we expand the scope of their work to include prioritizing grant proposals and reporting on achievements to the public, genetic fiction becomes a key background against which both scientists and the public imagine this work's repercussions.

Consider how Church invokes the transcendence of genetics in his monograph. While *Regenesis* (coauthored with Ed Regis) is praiseworthy for its direct, uniquely informed presentation of developments in a rapidly accelerating field—one in which the 2000 Bill Clinton/Tony Blair announcement of a completed draft of the human genome is now ancient history—the book's tone can hardly be called restrained. "But consider this modest proposal," the introduction invites. "What if it were possible to make human beings immune to all viruses, known or unknown, natural or artificial?" (8). A page later, discussing the possibility that some species' extinctions might be reversed, Church promises that "genomic technologies can actually allow us to raise the dead" (9). The spectacular imagery continues well beyond the attention-grabbing opening. Startling those who might prefer to retain distinctions between *Homo sapiens* and other life, *Regenesis* proclaims, "The interspecies barrier is falling as fast as the Berlin Wall did in 1989" (249). Church's method is to paint in bold colors and on the largest scales, as if echoing Flannery O'Connor's idea that "to the hard of hearing you shout, and for the almost-blind you draw large and startling figures" (34). The difference between her fictional use of the grotesque and Church's rhetoric, of course, is that a prominent scientist's hyperbole is more likely to be mistaken for a literal claim about reality, as if he really were planning to eliminate death or create superhumans without bothering to entertain the most basic bioethical questions.

I do not extract these moments in order to portray Church or any other geneticist as peddling biotech Kool-Aid. He is discussing real discoveries and remarkable applications that are realized or draw nearer every year, developments about which he has been admirably forthright. Although there have been moments in which biotech has seemed to lag behind its promises, as when the Human Genome Project revealed that gene-protein interactions are far more complex than previously imagined, there have also been periods of rapid acceleration, such as the recent scramble to take advantage of CRISPR technology as a delivery tool for gene editing.[1] I am noting simply that even for the most serious and respected scientists, the profound excitement of these developments makes metaphysical rhetoric about genetics very difficult to avoid, especially in a competitive funding environment. Genomic biologists and biotechnologists are surrounded by promises of a radical break, a revolution inaugurating a new era in personal medicine. There really are game-changing developments afoot, possibilities as barely glimpsed as the Internet's transformation of global communication and commerce was a few decades ago (especially if one wasn't a science fiction reader). In this sense, Church and other leading biologists should not be faulted overmuch for occasional overreaches; even when Craig Venter entitled a book *Life at the Speed of Light*, his enthusiasm was justified. Still, much can be lost in translation as scientists' metaphors fertilize the stories by which nonscientists make sense of things. We should consider carefully how these mutated narratives double back and colonize the research and applications that find private and public financing.

Beyond assessing the direct impact of actual biotechnologies, in other words, it is critical to assess the longer arc of their narrative manifestations, both for those who make these tools and for those who will rely on them. The most popular mythologies and the most sophisticated novelistic meditations shape attitudes long before interested parties order a direct-to-consumer personal genome testing kit or sit down with their doctors to consider a new form of gene editing. Sometimes these tales seem specifically designed to inform such decisions. But they often have other purposes entirely, as when they use genetic difference as a looser allegory for exploring racial or sexual identity. Either way, it is my view that some of the most penetrating reflections on the significance of genetics have needed to arrive via fiction simply because of the greater imaginative openness with which Western cultures usually approach fictional narratives as opposed to more traditionally pedagogical genres. In this sense, my approach to fiction about biotechnology builds on that of Susan Squier: "I understand fiction as a crucial site of permitted articulation for the desires driving these new biotechnologies: I am using this term broadly enough to encompass all linguistic play with *what might be*, all transgressions of the (socially constructed) boundary of fact, including the imaginative play of poetry. Fiction gives us access to the biomedical imaginary: the zone in which experiments are carried out in narrative, and the psychic investments of biomedicine are articulated" (17).

Focusing on the significance of genes rather than in-between entities like unused embryos, I aim to follow Squier in showing how novels, films, television shows, and graphic narratives about genetics are reshaping twenty-first-century conceptions of the self, the body, and various identity categories, sometimes in advance of the biotechnology they represent, sometimes in its wake. Critical of literature and science scholars' habit of "limit[ing] ourselves to using one field to gloss the other," Squier proposes instead to "us[e] them to unsettle not only each other but also their mutually opposed relation" (44). Indeed this book idealizes a similarly two-way street. Even as it draws most heavily on my disciplinary expertise as a critic of literature and film, it often warns against simplistically opposing the sciences and the humanities, instead looking for areas of potential cross-fertilization. In chapters 4 and 5 especially, *Editing the Soul* shows how the very form of contemporary fiction about genetics is mutating toward spiraling, self-referential structures that reflect an age growing ever more literate at the smallest of scales.

We'll wade into these theoretical waters in chapter 1. For now, to gain an initial sense of genetic fiction's diversity and its multivalent appeals to transcendence, let's step back and trace some further mutations in the biocultural metaphor utilized by Church's title, "Regenesis." This neologism speaks of a new creation, a remaking of the individual, but it also illustrates how the same metaphor can convey contradictory meanings. On one hand, it has become a favorite brand name in the cosmetics, body image, and pain management fields, where it is used predominantly to market processes for conforming to idealized images of the human body. ReGenesis Plastic Surgery and Skin Care Center in Colorado seeks "improvement, not perfection, in the way you look"; Regenesis Wellness Center in Arizona offers weight-loss training; a California company markets the ReGenesis Hair Enhancing System; New Jersey's ReGenesis LLC is ready to source materials for newly patented makeup and cleaning technologies; and Regenesis Skin Studio in North Carolina promises "beautiful, healthy, sexy skin." Sometimes the claims seem especially dubitable: Arizona's Regenesis Biomedical Inc. promotes a handheld electromagnetic therapy device that relieves pain by somehow "mediat[ing] cytokine and edema related genes" ("safe"), while the business plan of Texas's ReGenesis Behavioral Health promised physician access via videoconference, "just as if you were sitting together in the same room"—until the organization folded in 2016. My interest, though, is less in these ventures' credibility than in the longings to which they testify. They attract customers with an old lure that many still cannot resist: *You can start over, completely. Whatever your problem, we will wipe it from existence.*

On the other hand, there are also biotechnology enterprises that use the term *regenesis* to imagine longer-term, more gradual transformations. Environmental planning and restoration companies sometimes adopt the term to market solutions to relatively invisible issues created by an economy dependent on endless expansion. For example, Regenesis.com in California offers soil remediation, using biochemistry to handle messes created by groundwater pollutants from gas stations, dry cleaners, manufacturing, and other industries. Regenesis Management Group in Colorado and Regenesis Group Inc., a consortium of project consultants and educators with bases in three states, are also environmental mediation and planning companies. According to the latter, "Nature doesn't need our protection. She needs our *collaboration*." A company blog post explains the superiority of "regenerative development" ("Regenesis Group Inc.") to mere environmental restoration, arguing that the goal should not be to return a site to a prior, supposedly untouched condition but to redesign humanity's whole interaction with the surrounding ecology. This far more sustainable vision of regenesis points to interventions that might be overlooked on the surface but promise lasting benefits.

In the worlds of personal genetic testing and gene editing, which version of the metaphor will prevail? Will humanity's growing capacities to know and remake itself at the cellular level prove only the latest expressions of a consumer culture driven by fast profits and the temporary veneer of beauty? Or will emerging biotechnologies offer serious engagement with the deepest, sometimes invisible roots of individual and societal problems? I don't think the best genetic fiction is primarily or only interested in making predictions.[2] However, this book proceeds on the conviction that both optimistic and worrisome possibilities for regenesis are very real and are already occurring, with mixed results that can offer hope and caution in the same moment. Like instantaneous global communication and media, biotechnology is capable of proving both blessing and curse, with some impacts that are difficult to assess. The only further assumption I make is that the less individuals and societies understand these discoveries and applications, the more dangerous the consequences will be. Conversely, the more we grasp what such knowledge and tools actually can and cannot do, the greater their positive potential.

In other words, new biotechnologies are often fascinating, but this book is interested in more than the actual knowledge and devices and the bioethical questions they raise. *Editing the Soul* is most invested in showing how the promises of new biotech can make us overlook the more common cultural tools by which we shape the future every day: the stories we tell. We need to pay attention to not only genetics and its applications but also the cultural mythologies by which we interpret them, mythologies that routinely engage ultimate questions. This book will challenge various misperceptions about genes and genomes, but its deepest questions are about the complex, constructed visions of the self that contemporary genetic fictions explore. How are these works shifting our self-understanding of personhood, performing their own forms of conceptual engineering well in advance of the genetic devices they imagine? Might attention to their varying uses of genre conventions lead to better testing and intervention protocols? What perspective do they offer on the biological choices we have already made? Examining popular and avant-garde works alike, including literature, cinema, television, and graphic narrative branded as science fiction, *Editing the Soul* investigates a broad set of interactions between biotechnology and contemporary culture, encounters that necessitate changing many of our reading and viewing habits.

Three Forms of Genetic Fiction

"Science fiction," which encompasses the subgenre I call "genetic fiction," is a beautiful oxymoron. At its best, science fiction (SF) fuses logical appeals to demonstrable, repeatable evidence and imaginative yearning for the unknown, even the unknowable. Like other genres, it carries a long history of common motifs, from laser-wielding robots to interstellar spaceships, and those are off-putting for some readers. Still, even if one does not love the pulps or the Golden Age of Asimov, Clarke, and Heinlein, science fiction should be appreciated for its representations of nontraditional viewpoints and performances of open-ended thought experiments. Famously defined by Darko Suvin as the literature of cognitive estrangement, SF has gained cultural currency in this century for many reasons. Some artists and publishers have rebranded it as "speculative fiction" (a move that has sometimes been welcomed, sometimes not), while academia has increasingly rejected the elitism that once enforced harder boundaries between "high" and "low" culture. Respected mainstream authors and directors have increasingly utilized the genre's strategies and motifs, sparking scholarly references to "slipstream" or "outsider" science fiction.[3] Whatever one's interest in the genre's history, technological breakthroughs and cultural shifts have made the gap between science fiction and consensus reality seem to narrow in the early twenty-first century, and that has raised its interest level for many readers and viewers.

Given science fiction's ongoing evolutionary capacity, this book does not seek to add to the definitional debates. Convinced by John Rieder's essay "On Defining SF, or Not," I use the term broadly to describe a range of stories that are committed to making the strange seem familiar and the familiar strange. This sometimes occurs in fiction about outer space centuries into the future, with technical innovations ranging in plausibility from satellites to transporter beams, but it can also occur in stories that rely on a single alteration of historical events (e.g., the Allies losing World War II, as with Philip K. Dick's *The Man in the High Castle*). Science fiction might be colonialist and sexist, but it can also offer fresh new visions of peacemaking and healthy complications of gender roles. Edgar Rice Burroughs's *A Princess of Mars* and Ursula K. Le Guin's *The Left Hand of Darkness* feature radically different anthropologies, political theories, and attitudes toward gender, but both are SF. And in my broad definition, so are many elements of Ralph Ellison's *Invisible Man* (think of the hospital scene), Karen Tei Yamashita's *Tropic of Orange* (with its fusion of magical realism and technoculture), and Yann Martel's *Life of Pi* (from its carnivorous island to the overarching alternate-reality puzzle).

Likewise, the texts featured in this book cover a broad spectrum in their orientation to science, from those that use biological concepts and possibilities loosely to more technically detailed works. As genetic fiction, what they necessarily share is an interest in how increasing knowledge of genes and the genome (and epigenome and preteome) cultivates new understandings of what it means to be human and of how our species relates to the more-than-human world, whether the science fictional innovation—or "novum," as I call it henceforth—is a newly inexpensive testing capacity or a new tool for gene editing.[4] Historically, variations among more and less technical SF have sometimes spawned rigid "hard" versus "soft" divisions. But if we avoid treating such labels as exclusive, they can attune us to genetic fiction's distinct thematic emphases and already rich history of aesthetic combinations. To that end, I want to briefly explore three twenty-first-century stories that pick up the "regenesis" moniker in their own ways, whether overtly (in the first two cases) or more subtly (in the last). Representing three distinct forms of genetic fiction, these texts also highlight the book's interest in rethinking ties between gender and "technotranscendence," my term for a uniquely modern form of the sublime that combines religious and secular awe in the face of industrial and eventually digital achievements (a concept I will return to in chapter 1).

The three narratives ahead illustrate a more general pattern in how fiction evolves in relationship to specific discoveries and technologies. First, a new finding or tool is often treated in a relatively fantastic, heavily analogical form, sometimes using distant-future science fiction or superhero narratives to comment on contemporary circumstances. This does not mean the texts in question are "immature" or unworthy of close attention; while some might be less likely to reward close rereading, the key element is their requirement that readers or audiences suspend disbelief. Gradually, though, the most poignant science fictional novums tend to inspire more technically detailed, immediately plausible realism; in other words, these are texts that place increasing weight on the mimetic illusions they produce. Whether they are categorized with science fiction proper, such narrative mutations suggest that a given scientific or technological subfield has penetrated public consciousness more fully than in the past. Finally, the new knowledge or tool begins to inspire the most complex forms of metafiction, stories that increasingly blur the fantastic and the real. Here narratives do not just engage new findings or devices, nor are they primarily concerned with demonstrating their plausibility. Instead, they reflect on their own capabilities to shape our perceptions. In this way, tales that urge readers to think about technology gradually become their own technologies of thought. While they invest in transcendent language to widely varying degrees, such fictions regularly shape inherited mythologies into new forms, laying the groundwork for future attitudes toward science in the process.

Loosely applied, this progression might be used to organize a larger history of science fiction, from the more fantastic pulps and Golden Age works that dominated the early- to mid-twentieth century to the increasing realism and self-reflexivity of the later New Wave and cyberpunk. In the case of genetic fiction specifically, the pattern is somewhat more compressed, with a visible shift from genetic fantasy into genetic realism and then genetic metafiction over roughly the last half century. The distinctions between these modes are not unchanging, and I am less interested in their exact boundaries than in their value for identifying distinct narrative purposes and structures. They overlap at various points, with genetic realism growing out of fantasy and then metafiction emerging from each of the preceding strands rather than replacing one or the other. And as with science fiction as a whole, there are essential earlier texts—from *Frankenstein* (1818) to *The Island of Doctor Moreau* (1896) to *Brave New World* (1931)—whose influence are frequently acknowledged by their literary descendants.[5]

Without drawing absolute boundaries, then, this book is organized around three primary modes of genetic fiction. It begins with the 1960s and especially the 1970s, when popular speculation about reading and perhaps recombining genes surged ahead of scientists' actual knowledge and capabilities. In this period, genetic fantasies were most interested in science's allegorical value for engaging differences of class, race, gender, sexuality, and religion. Gradually, more biologically accurate treatments started to catch up, with works of genetic realism proliferating particularly during and after the Human Genome Project, from the 1990s into the 2000s. In this phase, the sometimes wild-eyed questions about homogeneity and heterogeneity raised by genetic fantasy became more nuanced and less capable of reduction to specific political positions. Finally, and quite recently, literature has started to seriously examine the self-reflexive potential of genetic metaphors, not just in the occasional winking aside, but in their entire organization. Genetic metafiction describes an increasing self-awareness in fiction about cloning, genome testing, and gene editing, one that blurs fantasy and realism and that reinvigorates discussions of ultimate questions about what it means to create and to be a person.

To reiterate, the emergence of genetic metafiction does not mean that genetic fantasy or genetic realism has disappeared. Blockbusters like *Jurassic World* (2015) continue to offer fantastic spectacles that say far more about the dangers of human hubris than about genetic biology. Meanwhile, as we will discuss later, the television show *Orphan Black* (2013–17) includes some fantastic elements, but it is much more committed to scientific plausibility and realistic character development. The trend I am identifying is not successive, in other words, but cumulative, and for this reason, the three modes are best understood in conversation with one another. Even if a reader's interest focuses on one of them, it is revealing to consider its relationship to the others. Most critically, we need to grasp the way each mode of genetic fiction emerges from its ancestor without rendering that mode extinct, an evolutionary pattern that sheds considerable light on how overblown rhetoric and naïve assumptions about biotechnology continue to dominate our news and social media, even when far more nuanced reflections exist simultaneously. The following recent examples of genetic fantasy, genetic realism, and genetic metafiction give a sense of how each uniquely plays on and critiques the distinctive form of transcendence commonly ascribed to contemporary biotechnology—one that is also strikingly dependent on cultural stereotypes about gender.

Genetic Fantasy and the *X-Men*

There might be no more triumphant visions of the genetic sublime than the opening credit sequences repeated across many *X-Men* films, the first of which appeared just three weeks after the 2000 Clinton/Blair press conference about the Human Genome Project. Accompanied by soaring soundtracks and cutting-edge CGI, viewers ride virtual rollercoasters into the inner machinery of the cell, a site full of electric explosions and fluorescent colors. While honoring the digital magicians and green-screen stars of the subsequent narrative, these sequences prime our imaginations for the miracles ahead. They also prepare us to overlook how little these films and the comics that inspire them are concerned with actual biology. As genetic fantasies, *X-Men* films and comics are more interested in using genetic exceptionality as an analogy with which to probe controversies around various categories of identity: gender, sexual preference, race, citizenship, and so on. The enduring debates of the *X-Men* universe are less about the effects of personal genome testing or new gene therapies than whether experiences of prejudice and injustice should inspire reform or spark revolution. Put bluntly, the franchise's central question is how it is possible to love one's enemy, particularly one exhibiting prejudice based on a simplistic biological determinism.

This allegorical significance does not make genetic fantasy irrelevant to those interested in the actual science. Genetic fantasy is the most common subcategory of genetic fiction by which the public encounters new possibilities for testing and modifying the human genome. Many more readers and viewers will encounter genetics via the *X-Men* than via George Church's book, no matter how many late-night interviews he does. So what do these superheroes teach us about how biological inheritance relates to individual traits? What kinds of responsibility do these narratives assign to genetics, environment, and personal agency? In a moment, I will flip through the pages of an *X-Men* comic, but for an even more familiar starting point, recall the film *X-Men 2* (2003) and its conflict between Ian McKellen's rebellious Erik/Magneto and Patrick Stewart's more restrained Charles Xavier/Professor X. This is something more than a stereotypical battle between "good guys" and "bad guys" that elevates one group above another. By confronting those whose means contradict their ends, *X-Men 2* epitomizes the franchise's emphasis that being a "mutant" or "other" does not make one automatically dangerous—or trustworthy.

This tension is often mapped racially, as Magneto recalls the revolutionary rhetoric and strategies of Malcolm X, and Professor X echoes the more patient but equally determined Martin Luther King Jr. Whether an audience recognizes the historical allegory, the lesson is that difference is good, that defeating hatred requires a balanced combination of assertiveness and restraint. In short, the film's diegetic science might be wildly implausible, with biological processes fast-forwarded to absurd speeds, but it deploys DNA in dynamic, coconstitutive relationships with larger cultural movements rather than rendering it as a simplistic, all-determining source of fate. There is a double reality here: the genetic specifics might be ludicrous, but the *X-Men* franchise is often deeply insightful about the ultimate inseparability of biology and culture. As Siddhartha Mukherjee emphasizes in his encyclopedic *The Gene: An Intimate History*, terms like *mutated* and *normal*, *sick* and *healthy*, or *diseased* and *enhanced* can only be defined in conversation with one another. Genes that are ordinary in one body or one environment might prove very poorly adapted to another.

Genetics and culture are also interwoven in an *X-Men* comic story arc that shares Church's title. Written and drawn by Jason Aaron and five other artists, *X-Men Regenesis* launched in 2011 with five issues entitled "Schism" and a special issue entitled "Regenesis" and then continued for nearly a year across several Marvel titles. Featuring a clash between two mutant groups led by Cyclops/Scott Summers and Wolverine/Logan, the narrative echoes tensions between Professor X and Magneto. This might seem only the most routine of marketing ploys, one congruent with the endless *Alien vs. Predator* and *Batman vs. Superman* sequels that seem to captivate Hollywood investors, but it also emphasizes one of genetic fantasy's strengths. Literature and cinema featuring new biotechnologies regularly use these motifs less for their referential ties to actual-world possibilities than as analogies by which to engage immediate sociopolitical circumstances. Genetic fantasy might employ actual biological terms and concepts, and it might even convey explicit interest in the future of biotechnology corporations—as in the film series' most recent entry, *Logan* (2017)—but in many ways, its biological specifics remain window dressing. These narratives' deepest concerns are with already immediate social challenges and identity problems, and the real danger lies in audiences' mistaking genetic fantasy's biological rhetoric and imagery for factual representations of actual biology.

In the comic *X-Men Regenesis*, after yet another attack on mutants' right to exist, the central question is where readers decide their beliefs align. Do they join Cyclops in the view that the genetic differences between mutants and nonmutants are insurmountable and warfare is inevitable, or do they agree with Wolverine in the insistence that reform and integration remain possible? In echoing the history of race relations and gay rights, the debate gestures to the potential for genetic engineering to exacerbate gaps between social classes.[6] Yet the underlying issue is how to deal with difference *now*: whether to retreat into homogenous enclaves or to keep taking the risks required for integration. Examined closely, the panels of this comic suggest not only how little genetic fantasy has to do with laboratory genetics but also how indelibly it continues to shape public perceptions of that science. A guest writer for this kind of *X-Men* comic needs to know nothing about today's actual personal genome testing or modification possibilities—only how readers expect the metaphor of the gene to function in the *X-Men* universe. That purpose is simple: the "X gene" drives a wedge between ordinary humans and simultaneously rejected and worshipped super/subhuman characters, pressuring the latter to decide whether they will use their power for the sake of vengeance or reconciliation.

Even a fairly mundane, actionless panel like that in the concluding pages of the first issue of *Schism* is telling (fig. 1). As tensions mount between mutants and "normal" humans, a group of angry whiz kids are using enormous sentinel robots to raise international havoc. Their warmongering has an immediate effect on immigration and border control policy, a biopolitical connection that grows even more common in genetic realism. For our purposes, the panel's main significance is its illustration of how independent genetic rhetoric can be from actual biology. Here the novum of "gene-sensitive land mines" is a farfetched proposition with no more real-world significance than Cyclops's reference to a "functioning X-gene." Contemporary genetics has made clear that only rare traits or diseases can be linked to a single gene, so reading literally, it would be preposterous to suggest that one gene might control the innumerable variety of superpowers that the comic's heroes and antiheroes exhibit. To then posit that such a gene might be detectable from belowground—instantaneously and with no means of analyzing a specimen—only drives the text further from plausibility.

Fig. 1 *X-Men Schism* #1 (2011) illustrates the gap between the term *gene* in genetic fantasy and the actual science on which it is based.

Yet this realization does not render *X-Men Regenesis* irrelevant to actual genetic science. Even when genetic fantasies have little to do with actual biology, they often powerfully shape science's public reception. Simply by *seeming* to find mystical powers in mutations, they affect attitudes toward genetic biology's real capabilities. In cases like the cliché-ridden horror flick *Splice* (2009), the audience has little opportunity to empathize with the technomonster, and the effect is only to heighten unnecessary fears, but in examples like this *X-Men* comic, the impact is more complex. Despite its biological baselessness, this genetic fantasy helpfully challenges absolute divisions between tiers of humanity, instead nurturing the self-images of those whose abilities are less quickly recognized by conventional educational measurements. Dismissing such texts as exercises in adolescent escapism would overlook how even their supernatural excesses call attention to our inflated biotechnological expectations. Genetic fantasies like *X-Men Regenesis* suggest that as long as biotech is conflated with magic, our culture is unlikely to grasp its actual costs and benefits. Just as the horror elicited by Frankenstein's creature depends on a terrorized population's ignorance of his perspective, the comic interrogates the rapidity with which humans take difference as threat. Treating by analogy those oppressed on the basis of nationality, race, class, gender, sexual orientation, and physical ability, *X-Men* stories cultivate both individual and group agency. The "genetic mutations" are a bridge: the true goal is to imagine redistribution of power to the disenfranchised.

With that in mind, we should pay particularly close attention to the religious elements of these *Regenesis* characters. In this story arc, villain Kade Kilgore's narcissism is attracted to the "impeccable genetic purity" of the Hellfire Club and its eugenic aspirations, but the text encourages readers to root for the richer hybridity and mixed allegiances found among the X-Men. Is this only another instance of a secular plea for tolerance against the militancy of religious fundamentalism? The page spread in figure 2 is a telling microcosm of the comic's complex answer, one it achieves via simultaneous utilization of and resistance to technotranscendence. Here the X-Men—or rather, X-Women—are juxtaposed against gargantuan robotic warriors loosed on a group of jihadis. While exposing the bigots' coercive appeals to divine authority, the comic renders a militant Islamism as inferior to these genetically evolved goddesses, who easily defeat mere steel and programming. From this angle, the comic might appear to be elevating a general humanistic discourse of equal rights above religious hatred. However, that reading overlooks the fact that this is a three-sided battle, that the X-Women are actually rescuing religious fundamentalists from the cold militarism driving their robot attackers. Instead of identifying the superheroes as secular opposites to the jihadis, the superheroes actually represent another possibility entirely. As the young X-(Wo)Man rescuing the sexist militant in the bottom right panel sardonically observes, "The fact that we're all women, well, I'm sure that's just a coincidence." Then she rubs in the irony: "Did I happen to mention I'm also Jewish?" The combination of religion and gender here is also no coincidence.

What *X-Men Regenesis* points to is the need for a third category for thinking about the religious and the secular, much like the urgency of reaching beyond easy male-female binaries. I will argue more fully for that postsecular approach a bit later on—not only to genetic fiction but to literature and cinema about science more broadly. Here I will say only that while postsecular theory is critical of modern secularisms, including those promoting the technological sublime, my deployment should *not* be misunderstood as a covert plea for simply returning to traditional religiosity or abandoning distinctions between religious and political authority. As in the case of the Jewish X-Woman, postsecular thinking is sometimes attractive to interpreters with theological backgrounds, but it can be just as useful for nontheists. It demonstrates how the religious and the secular are only definable in relation to one another and only in relation to specific cultures. What counts as "secular" in rural areas of the US South or Midwest is not what counts as "secular" in Tokyo or Shanghai. Postsecular thinking also tends to be nonhierarchical, and that might be why in fiction, it often critiques patterns of sexism and fundamentalism simultaneously. Both forms of abuse

Fig. 2 *X-Men Regenesis* (*Schism* #2, 2011) uses genetics to illuminate other cultural rifts.

depend on authority that is enforced rather than earned, and when complex characters find the courage to challenge patriarchy, religious confrontations often await—and vice versa.

I have concentrated so far on genetic fantasy's potential to inspire nuanced understandings of genetics, but that does not mean it cannot also convey shortsighted, ill-informed assumptions. Because most audiences' familiarity with actual genetic biology is minimal, it is easy to read the metaphors of genetic fantasy simplistically, so that they substitute for knowledge of the real thing. In truth, laboratory genetics is widely distinct from the determinism pervading "astrological" genetics, just as it diverges from "genetic dismissivism," my term for the readiness of some to ignore or downplay this field's expanding value. Readers with less scientific knowledge might naturally find genetic fantasy most attractive, as the absence of technical scientific details means they will rarely find themselves at a loss. Conversely, readers with stronger scientific backgrounds might discover surprisingly valuable elements of genetic fantasy, like its ability to foreground confusions that arise often in conversations between biologists and nonscientists. I hope this discussion will encourage readers on both sides of the spectrum to consider alternate perspectives.

Genetic Realism on Canadian TV

While *X-Men* comics date to the early 1960s and were preceded by other instances of genetic fantasy, a second subcategory of genetic fiction only began to appear as public awareness of genetic engineering expanded in the very late twentieth and early twenty-first centuries.[7] Like genetic fantasy, its attitudes toward biotechnology's blessings and curses vary. In some instances, genetic realism adopts a cautionary posture, with all the concerns about "scientists playing God" one might expect. Yet the category also features many more affirmative texts—such as another *ReGenesis*, a Canadian television show (2004–8) that imagines Canada, the United States, and Mexico overseeing public health, monitoring new biotechnologies, and fighting disease outbreaks with a single transcontinental research agency. The show won an enduring viewership through its unusually informed depictions of the lengthy, sometimes fruitless processes of laboratory labor and of research scientists as holistic, internally conflicted characters. Even when utilizing accelerated presentations of complex cellular processes and medical decision making, *ReGenesis*'s portrayal of a North American Biotechnological Advisory Commission (NorBAC) makes it one of the most intensely pedagogical of biological science fictions ever to hit the small screen. Imagine a *CSI* in which there is not only thorough scientific research in the mise-en-scène, but in which the mechanism of a specific gene's mutation is integral to the plot. Beyond genetic fantasy's allegorical questions about social justice and identity, the series epitomizes genetic realism's more precise attention to biological possibilities. It aptly represents genetic realism's capacity to resist the entreaties of technotranscendence, instead coaxing audiences to soberly consider biotechnology's near-term opportunities and consequences.

For a snapshot of the show's investment in public education—and one that shares the *X-Men*'s resistance to patriarchy—consider a storyline from the show's first season.[8] NorBAC CEO Caroline Morrison receives a request from a US congresswoman to provide an objective outside ruling for an innovative cancer gene therapy trial that has been temporarily suspended. The therapy, based on then current knowledge and intervention strategies, is for pediatric acute myeloid leukemia (AML) patients for whom conventional treatments like chemotherapy have failed. Viewers swiftly learn that one of the first three gene therapy recipients died within several hours, whereas the other two children experienced remarkable recoveries. As a favor to a political ally—herself responding to a colleague's plea to help his AML-ridden child—Morrison takes her lead scientist, Dr. Sandstrom, to investigate. It turns out that the objections about the trial, led by a Dr. Booker, have come from a female biologist who finds the gene editing trial's risk level intolerable. Unfortunately, the whistleblower, Dr. Foley, has some awkward romantic history with Sandstrom, which only contributes to the rage she feels at his eventual decision to let the trial continue, even after a second child dies.[9]

This brief plot line highlights two attributes of genetic realism that grow increasingly evident as we approach the present. First, rather than bolstering biological determinism, *ReGenesis* portrays biology and culture as endlessly redefining each other. Unlike genetic fantasy, genetic realism is often revelatory of the complexity of the individual and collective choices that drive biotech investment, research, and interventions. To that end, one of the CBC show's chief concerns is with enduring assumptions that female scientists are especially prone to subjective, emotional interpretations of evidence. It is no surprise that the biologist who objects to Dr. Booker's trial is a woman or that Sandstrom shows little inclination to separate Dr. Foley's claims from his memories of their youthful fling. Booker accuses her of "jumping the gun, making decisions based on emotion, not science," and Sandstrom nearly accepts this judgment outright ("Wouldn't be the first time," he notes). At first glance, in fact, the show might seem complicit in this sexism. The last time Foley appears onscreen, she shows up uninvited at a party at Sandstrom's home to confront him with news that, after he let the experimental treatments continue, a second child had died. The fact that six other patients are showing signs of recovery is irrelevant; all she can see is that "you signed that boy's death sentence." The episode might appear to condone gender hierarchies in science, given how extremely this female accuser behaves.

Instead, *ReGenesis* slowly unpacks a disturbing tendency within cultures of medicine to render epistemology a matter of gender. To recognize how the show links sexism with objectivism note Sandstrom's interactions with another female character. Eventually offering a different vision of gender roles, the sequence exemplifies genetic realism's interest in rendering biological details in terms that are compelling for nonscientists while also revealing how science sometimes places researchers in positions akin to clerics or even gods, expected to decide who will and will not receive a chance to live. Before things get better, they get worse: when Sandstrom and Morrison sit down in Booker's office, his side of the story quickly shifts into insider jargon, and via close-ups on his lips and her reactions, viewers quickly recognize the sexist dynamic:

> Booker: "See, cancer cells multiply at an extremely high rate. We deliver a simple cold virus that makes an RNAi. [Condescending aside to Morrison] RNAi: it's a new method for eliminating proteins like AML-6."
>
> Sandstrom [intrigued]: "And the RNAi kills the mutant protein in the cell and cell growth is arrested."
>
> Booker: "We're tricking them into committing mass suicide." [Double screen appears showing Sandstrom's enthusiastic reaction and the virus working in the cancer cells.]
>
> Sandstrom: "Apoptosis, wow."
>
> Booker [conspiratorial smile, warming to his subject and entering full genius mode]: "Exactly. The RNAi is encoded with an EPV-derived plasmid under a POLE 3 expression. Now the plasmid is targeting the tumor cells in a viral vector harboring a humanized anti-CD33 binding protein."
>
> Sandstrom: "Because CD33 is highly expressed in AML cells."
>
> Booker: "That's it. See, it's a triple-barrel approach. We attack the specific molecular defect in the AML cell, we induce apoptosis, and we can even ensure a higher specificity by using the CD-33 binding protein."
>
> Sandstrom: "That's amazing. Because of one death, this treatment is going nowhere?"
>
> Booker: "See why I'm so . . ."
>
> Sandstrom: "Pissed off?"
>
> Booker: "Yeah."

The gene therapy is brilliant, but the scene exposes the dynamic by which a particular style of heteronormative masculinity sometimes attempts to leave others behind. There's an obliviousness to Booker's and Sandstrom's glee that aligns them conspiratorially, shoving aside the bioethical concerns that prompted the visit and negating Morrison's presence. All they see is the beauty of the puzzle and its promises of a solution untarnished by subjectivity, emotion, or ambiguity. *ReGenesis* appears to be relying on a common set piece for medical thrillers: two (usually male) characters spout a litany of biological abstractions, and the audience enthuses, "Wow, those guys are smart!"

But like the best genetic realism, the Canadian TV show doesn't stop with erasing the female observer. The conversation picks up outside, where Morrison demands further explanation. This second dialogue, which begins with brief sexual innuendo, makes clear that Morrison is onto Sandstrom's sexist tricks—and he knows it. For all its initial playfulness, the conversation evokes a mutual respect that increasingly typifies the characters' relationship:

> Sandstrom: "If I had two samples of the same virus and I put them in ideal conditions, right, turn the lights down low, put on a little Barry White—" [flips up his middle fingers on both hands, wiggling them as they approach each other and try to link]
> Morrison: "I got it—"
> Sandstrom: "I'd never get a third virus because viruses don't reproduce that way."
> Morrison: "They're asexual."
> Sandstrom: "Ah, you've been doing a little reading, eh?"
> Morrison [sighing, but persistently]: "Go on . . ."
> Sandstrom: "So, in order for a virus to reproduce, they've evolved these neat little tricks of breaking into a host cell, right, killing the boss [machine gun impression], taking over—[stereotypical bad guy voice] 'OK, you guys are working for me now'—pretty soon you've got all these nasty little viruses out there in the neighborhood shot to hell."
> Morrison: "OK . . ."
> Sandstrom: "OK, so in gene therapy, you take a virus and you alter it. You remove all the nasty bits and replace them with something good. And now you've got the best delivery system Mother Nature has to offer. Because if you inject a few trillion of these nicely altered little viruses into a patient, the viruses will just do what comes naturally, and they will break into the patient's cell—"
> Morrison: "And kind of infect them with health?"
> Sandstrom: "Pretty amazing, eh?"
> Morrison: "Yeah . . . it is."

In contrast to Sandstrom's fight with Foley, the exchange pulls these characters together in a shared awe at the intricacies of cellular evolution, starting to contradict earlier suggestions that "women are too emotional." Instead, as the show increasingly indicates, the more fully we understand scientific nuance, the more thoroughly we should all be astonished.

Not just for nerds, then, genetic realism is for everyone. In fact, *ReGenesis*'s producers so heavily prioritized audience education that they partnered with the organization then known as the Ontario Genomics Institute (OGI; today it is called Ontario Genomics) to produce materials aligning each episode's biological details with current knowledge (fig. 3). Not surprisingly, given this partnership, the show generally treats its scientific protagonists sympathetically. Like *X-Men* narratives in which mutants disagree about how to respond to persecution, it avoids easy character stereotypes and emphasizes the inseparability of genetics and culture, acknowledging that money, politics, and personal prestige play major roles in shaping scientific discoveries and biomedical innovations. In the episodes examined here, it would have been easy to paint Booker, the gene therapy project's principal investigator, as an entirely selfish villain; Foley could have then played the courageous maternal figure stepping in to save the juvenile "victims." Instead, both scientists emerge as complex figures, despite limited screen time over just two episodes.

Two structural elements of these *ReGenesis* episodes also point to genetic realism's potential. First, the bioethical mess they dramatize is compounded by the fact that we never see the final outcome of this genetic editing. Whereas the television show *ER* (1994–2009) drew comparisons to soap operas in its efforts to resolve minor plot lines, *ReGenesis*'s conclusions are more comfortable with uncertainty. Moreover, this example of genetic realism does not allow its storylines to subsist in a vacuum. In these episodes, we also watch a gay man seek an experimental treatment that relies on nonprogressers, people who have HIV but never get AIDS. When a NorBAC researcher (his former lover) helps him gain access, we see the treatment injected into the sick man's bloodstream but not the result. For all its biological specificity, the show is less interested in promoting a specific bioethical position than in underlining the subjective risk taking involved in any application of cutting-edge science, no matter how strict the human subject protocols. If there is an ideological claim here, it is only that the soul deserves dignity no matter the body's condition, that whether new biotechnologies end up cultivating or destroying relationships depends on both individual choices and social policies.

In sum, when the show's cautionary tagline inquires "Have We Gone Too Far?," *ReGenesis* typifies genetic realism's reliance on the interrogative mode. Avoiding oversimplifications of biological uncertainties, the series remains

ReGenesis | **Episode 1.6: The Trials**

NorBAC reviews an experimental gene therapy treatment being used on children with leukemia. Just as they are about to declare the cancer therapy safe, another child dies.

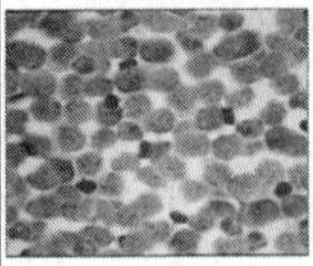

Blood cells
Leukemia is a cancer of the blood cells that causes too many immature white blood cells to circulate in the bloodstream.

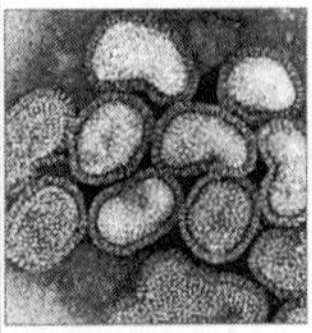

Influenza virus
Viruses are commonly used as vectors to deliver gene therapy to the affected cells in a patient.

Is it possible to use gene therapy to treat leukemia?

What is acute myelogenous leukemia?
Leukemia is a cancer of blood cells. Acute myelogenous leukemia (AML) is a type of leukemia that leads to too many immature white blood cells in the bone marrow and blood. AML affects both children and adults.

Blood cells are part of the immune system. They develop from stem cells. These stem cells can develop into two types of cells, lymphoid stem cells, which develop into white blood cells, and myeloid stem cells, which develop into both red and white blood cells. AML is a cancer of the myeloid cell type; the myeloid stem cells fail to develop into mature white blood cells, and can also develop into abnormal red blood cells. These leukemia cells build up and can prevent healthy cells in the immune system from fighting infection. Leukemia cells can spread to other parts of the body including the brain, spinal cord, skin and gums.

What are common treatments for leukemia?
The two most common treatments for leukemia are chemotherapy and radiation therapy. Chemotherapy is used to help treat cancer because these drugs target and kill fast-growing cells. Radiation therapy involves focusing high-energy rays to kill leukemia cells, in the spleen, brain or other areas.

Because these treatments kill both disease cells and healthy cells, they often leave the patient's immune system weakened and unable to fight off infections. Stem cell transplantation may be used in combination with chemotherapy or radiation therapy. Prior to receiving chemotherapy or radiation therapy, some of the patient's own blood stem cells are removed (in bone marrow) and grown in the lab. After the patient receives therapy, the patient's own blood stem cells are transplanted back to help re-populate the weakened immune system with healthy cells.

How does gene therapy work to treat disease?
Many diseases including cancer occur when genes – which contain instructions for making proteins – become mutated. A mutated gene can result in a mutated protein that doesn't work properly, or sometimes no protein at all. Depending on the gene, this could lead to serious problems and disease. Mutated genes can be inherited, or they can arise from exposure to chemicals or mutagens; smoking and sun exposure are both known to cause DNA damage that can lead to mutations. What if we could "fix" the mutated genes that cause the cells to be cancerous? Gene therapy aims to restore the normal functions of cells by introducing a "normal" copy of a gene.

For many blood and immune system related diseases such as leukemia and lymphoma, gene therapies could either target the disease cells to restore normal function, or the immune cells to enhance their ability to fight the disease. Generally, disease cells are removed from the patient, treated with gene therapy to introduce a normal copy of a gene into each cell, and then those treated cells are put back into the patient. The gene therapy itself involves engineering a virus to include the normal human gene. The cells removed from the patient are infected with that engineered virus. When the virus infects the cell, it inserts the gene somewhere into a chromosome in the cell – the cell now contains a normal copy of that gene. This normal gene then makes a normal protein, which can help restore normal cell function.

12

Fig. 3 Page one of *ReGenesis* public education pamphlet, season one, episode six.

conscious that no science occurs without the inspiration and mediation of specific cultural values. It is not just a matter of following clear, automatic moral precepts or official policies. Instead, individual and societal choices to utilize or abstain from particular tools and methods are inevitably messy, nonobvious decisions. Though sometimes more cautious about biotechnological advances than other entries into our cultural dialogue about new biotechnologies, genetic realism asks informed, penetrating questions about present and near-future testing and intervention capacities. Far from leaving behind the questions about sex, gender, race, and religion raised via the allegorical thrills of genetic fantasy, this subcategory of genetic fiction often uses its deeper research and its greater comfort with ambiguity to lean on them all the harder.

Genetic Metafiction as Origin Myth

As stories in the genome age have become more sophisticated, the lines between realism and fantasy have continued to blur, and the narratives have grown increasingly self-reflexive, yielding a third, nascent subcategory I call genetic metafiction. These stories can emerge from genetic fiction's more fantastic or more realistic strand, and this section briefly illustrates the latter possibility.[10] Richard Powers's 2012 short story "Genie" is not quite entitled "regenesis," but it deserves that term's associations with remaking humanity as much as Church's book, the *X-Men,* or the Canadian TV series. The story interrogates its overtly realistic, hyperinformed approach to genetic biology by placing it within what is ultimately an alien "first-contact" narrative. While spotlighting the metaphysical appeal of current and near-future biotechnologies, the story zooms out to ask the deepest of questions about the establishment of knowledge. In the process, "Genie" comments on its own capacity to illuminate the processes by which human beings make meaning, especially in an era when human devastation of other terrestrial life has grown exponentially. Also featuring an unconventional romance, Powers's tale epitomizes genetic metafiction's interwoven, self-conscious questions about epistemology, metaphysics, and the significance of human relationships.

In "Genie," after swiping biological material from a geothermal pool at Yellowstone, a young postdoc uncovers evidence of a primordial genetic order, a possibility she finds immensely disturbing. It appears to Anca that instead of pointing to the gradual pattern of natural selection, her sample evidences the intentional signature of a "watchmaker." It's not that Powers's story has suddenly aligned itself with the contemporary intelligent design movement; rather, his story displays an integrity that reaches far beyond many efforts to defend science against anti-intellectualism. Instead of simply aligning science with religion or pitting these forces against each other, "Genie" becomes an eloquent evocation of genetic biology's enormous vistas and our unpreparedness to deal with its findings. In structure as well as in explicit themes, the story attests to genetic metafiction's fascination with the most poignant questions about love and personal meaning. Most provocatively, it proposes that genuinely loving any other requires a soul ready to love *all* others, whether human, animal, or extraterrestrial. This might seem strangely mystical content for a story by one of contemporary literature's most scientifically educated practitioners, but it clearly shows that if we pigeonhole phenomena as exclusively or necessarily scientific or cultural, spiritual or material, holy or profane, we adopt a binary two-dimensionality that artificially separates reason from intuition, knowledge from faith, wisdom from emotion. Some of the hardest of biological science fictions and the darkest of contemporary realistic novels and films now depend on the friction created between physical and metaphysical thinking.

"Genie" offers several compelling demonstrations of genetic metafiction's potential. No aliens physically materialize in Powers's story, yet the tale's first climax arrives with Anca's realization that her stolen spoonful of primeval soup was genetically "autographed" four billion years earlier. Featuring far too many nonrandom sequences to impute to chance, the sample testifies that someone far more scientifically advanced than contemporary humans visited Earth when its only life-forms were microbial. This is a difficult pill for Anca to swallow, as everything in her background resists such a conclusion. Alongside her on-again-off-again lover Warren, who contributes not just coding skills and statistical analysis but a humility cultivated by reading Thomas Merton, she finds herself "wrestling with Paley's pocket watch on the hillside, all over again, and the suggestion of a watchmaker at work panicked her." Desperately casting around for alternative explanations, she convinces herself that the string of apparently nonrandom data might have resulted from natural selection. But Warren points out her explanation's insufficiency, and "logic, in all its indifference to sense, said that he was right." On its face, it could seem that Powers, easily one of the most scientifically literate novelists in the history of American literature, had suddenly developed a penchant for magic.

A more nuanced interpretation is that Powers's story is a stirring demonstration of the fact that those who take science seriously must be willing to go wherever the evidence leads, even if that seems to play into the hands of antiscientific forces. It is entirely possible for everything known about science to be correct and for there to remain influences beyond our capacity to detect. Evolution by natural selection, the universe's origins in a big bang, and so on could all prove congruent with some larger reality: these could be the means, the governing patterns, by which humanity interacts with entities beyond its capacity to verify. But such possibilities are outside the bounds of proof; we can only guess, take risks, come back with new stories, and repeat. Thus "Genie" is a thought experiment about what the most dedicated efforts to probe "theories of everything" should reveal. It is simultaneously about the potential of genetic science to uncover new realities about our species and about the capacity of genetic fiction to take on metaphysical meanings that go well beyond routine laboratory findings. In a short story that constantly references the strangeness of the weather and the silent but inexorable shifts of climate change, Powers imagines a persistent love that looks human frailty and insufficiency in the face and goes on nonetheless. It confronts evidence of the artificial constructions of human souls and societies and, without either triumphalism or despair, revalues them within a bigger vision of life's expanse and its detail.

"Genetic metafiction," therefore, involves fusions of not only realism with fantasy but also storytelling momentum with self-reflexivity and sensory data with ultimate questions. Hence Powers's story is devoted to the material and the testable but open to radical paradigm shifts. Rather than endorsing passivity, it embraces an active, determined searching that encompasses biological methods as well as cosmological inquiries. Anca's last name is Jaeger, derived from the German for "hunter," and readers are invited to join her and her ex in transgressing conventional boundaries between religion and science. Indeed she "live[s] in the *faith* of a coming great *discovery*" (emphasis added)—not in the certainty of revelation or the self-satisfaction of an all-sufficient empiricism. Assisted by the yin of Warren's patience, Anca discovers that their microbial sequence is not a random collection of A, T, C, and G nucleotides but encoded music. In ways they only begin to grasp, it harmonizes the microscopic with the telescopic.

Two metafictional moments in "Genie" are particularly telling for the significance of genetic metafiction. First, the narrator comments of the ex-lovers' breakthrough, "*The world's fullness is not made but found*" (emphasis in original). This is a quotation of Richard Wilbur's poem about the wedding feast at Cana described in John's gospel, and it suggests that life's miracle is inextricable from ordinary materiality. Wine begins as grapes, yeast, water; the "miracle" is finger-snapped into existence not in the style of a "genie" but in the acceleration of normal temporality. Second, the story suggests that readers might have to accept, as per Wittgenstein, that "if a lion could talk, we wouldn't be able to understand him"—a problem to which this book's coda will return. For now, we need only observe Warren taping Anca's base-pair sequences above a picture of the Lions' Court in the Alhambra. Just as that medieval Spanish Islamic holy site figures the meeting of rivers in a mythological paradise, "Genie" makes the pool at Yellowstone into a new American Eden, with Warren and Anca as its returning Adam and Eve. In stealing fruit from this wilderness garden, they find a new knowledge for which they are entirely unprepared. In that sense, Powers's story points to missions often articulated by genetic metafiction—to ready us for what is coming and to help us grasp what has already arrived.

In the pages ahead, chapter 1 provides two kinds of primers that should benefit many readers. Because this book is designed to engage audiences with diverse biological backgrounds, the chapter begins by detailing key differences between common cultural perceptions of genetics and the actual laboratory science. A gene is not nearly so simple as it is often represented, nor do genomes yet reveal so much as popular accounts sometimes suggest. Without attempting to substitute for a growing array of cogent introductions to genetics available to nonspecialists, I offer a brief primer that dispenses with some of the most common misunderstandings. Chapter 1 then turns to an inventory of key humanistic resources for the book, highlighting contributions from literary studies especially. Here we begin to see how genetic fiction engages long-standing epistemological questions, and I explain how and why this book takes up concepts of transcendence that might seem to belong entirely to religious studies. Building on insights from Michael Bérubé, Donna Haraway, N. Katherine Hayles, Michael Kaufmann, Robert Markley, José van Dijck, and Priscilla Wald, among others, the chapter concludes by articulating the value of postsecular theory to science studies generally and to my argument in particular.

Chapter 2 then dives into genetic fantasy by focusing on pre–Human Genome Project stories about human cloning. While almost universally condemned by bioethicists, this form of bioengineering has routinely transfixed readers and viewers of popular fiction and film, not least because of the provocative questions it raises about individual uniqueness. With new examples appearing regularly across the last half century, cloning tales provide an ideal opportunity to survey genetic fantasy's long development and to see how heavily it relies on the technological sublime. Noting these stories' enduring prevalence in blockbuster cinema, the chapter centers on their novelistic development in the years immediately before and after the pivotal 1975 Asilomar Conference on Recombinant DNA. While many twentieth-century cloning novels reflected astrological or hyperdeterministic genetic rhetoric, I devote the most space to feminist science fiction that pushed back. Routinely employing theological language and sometimes hinting at the imbrications of genes and environment that would grow more prominent in the next century, these works point to the need for a postsecular approach to biotechnology and its metaphors. In looks at fiction by Ursula K. Le Guin ("Nine Lives," 1968), Pamela Sargent (*Cloned Lives*, 1976), and Octavia Butler (*Xenogenesis/Lilith's Brood*, 1987–89), we uncover the process by which the "Carbon-Copy Clone Catastrophe" came to produce deeper insights than one might expect of this outlandish masterplot. Written before the leaps in knowledge that came via the Human Genome Project, these works were already interrogating the objectivism, scientism, and sexism that continue to beset much discourse around genetics. To that end, they envisioned radical, even queer manifestations of love and sexuality that might more fully integrate concepts of biology and culture. Emphasizing genetic fantasy's ongoing relevance, the chapter concludes with a story that initially appears heavily realistic but ultimately offers its own take on genetic fantasy, Duncan Jones's low-budget indie film *Moon* (2009).

Testifying to growing fears about how biometrics and other forms of biopower might police human freedom, the key texts in chapter 3 indicate how often early twenty-first-century realist literature has invested in science fictional tropes and questions. Like their more fantastic cousins, these "slipstream" works build connections among genetic technologies and sexuality, race, and other categories of identity, but they emphasize more heavily the roles of cultural forces, sheer accident, and individual agency in tempering genetic determinism. Before analyzing the literary realism at the center of this transition, I begin with a look at the very recent BBC America television series *Orphan Black*. This wonderfully strange chimera of alternate history and unusually informed biotechnological speculation straddles the shifting boundary between fantasy and realism, but when set against *Moon*, the accuracy of its genetic biology and its interest in character interiority pulls it closer to realism. The chapter then works backward toward the emergence of genetic realism. Adapting earlier cloning fantasies, Kazuo Ishiguro's alternate-history novel *Never Let Me Go* (2005) pursues a subjective attention to the ordinary that invites readers to recognize more fully than ever before that in a very real sense, we are all clones. Two of the most visible early novels of genetic realism, Zadie Smith's *White Teeth* (2000) and Jeffrey Eugenides's *Middlesex* (2002), refigure genetic science in terms of international immigration. Composed and set on opposite sides of the Atlantic, Smith's and Eugenides's uncannily parallel novels offer intergenerational tales that reflect many of the new century's anxieties about genetic inheritance. Beyond questions of race and nationality, issues of gender and sexuality are just as prominent here as in genetic fantasy. The im/materiality of the soul and the freedom to love are again prominent questions, but genetic realism addresses them with more specific attention to increases in scientific knowledge and to biotechnological options available to individuals and communities in the present and very near future.

With chapters 4 and 5, *Editing the Soul* concentrates on recent works that not only shift their audiences' visions of genetics but also self-reflexively redefine the relationships between science and fiction. First is an examination of the genetic metafiction at work in *Y: The Last Man*, a comic that ran from 2002 to 2008. This zany but poignant tale of the near disappearance of males across species remediates a literary last man/all-female society tradition that runs from Mary Shelley through Charlotte Perkins Gilman and Joanna Russ, in this case tying the Y-chromosome disaster to the act of human cloning. *Y: The Last Man* draws heavily on genetic fantasy; its conclusion proposes that a self-protective natural intelligence binds together the Earth, so that the moment a male cloned child emerged from the womb, nearly all the traditionally reproduced males on the planet had to die. Alongside this outlandish premise, however, the narrative regularly mocks its own plot and form, gesturing to the act of storytelling as its own form of cloning and questioning common assumptions about the relationships among biotechnology, gender, and religion. For these reasons, it is a valuable introduction to the capacities of genetic metafiction, one that reemphasizes genetic fiction's broader concerns with shifting conceptions of love and the soul.

The payoff of a postsecular approach is even more apparent for interpreting Margaret Atwood's *MaddAddam* trilogy (*Oryx and Crake*, 2003; *The Year of the Flood*, 2009; and *MaddAddam*, 2013). Currently being adapted for television by Darren Aronofsky, Atwood's novels combine her trademark black humor with serious reflections on the genetics of both humans and other species. The *MaddAddam* trilogy attends simultaneously to the genome's biological reality and its metaphorical significance, demonstrating why we can afford neither an overcautionary Ludditism nor an insufficiently suspicious technophilia, neither a kneejerk genetic determinism nor a passive genetic dismissivism. Instead, the "serpent wisdom" of Atwood's novels refreshes biblical motifs by building sympathy for her twenty-first-century Doctor Moreau and his creatures, joining *Orphan Black* and *Y: The Last Man* in twisting the Eden story to offer a sober, uncertain hope for what lies ahead. Most significantly, the trilogy translates the concerns of late twentieth-century cloning novels for a post–Human Genome Project era, exposing how worries about biotechnology continue to reflect larger, more general anxieties about both homogeneity and heterogeneity. In Atwood's storyworld, Earth's population has proliferated well beyond current totals (seven billion as of 2011 or 2012), and with all their fears of difference, human beings are growing deathly afraid that individual uniqueness is no longer possible. This fear cultivates a genetic determinism that turns simplistic and absolute when it needs to become more nuanced and humble. By extrapolating from actual biotech developments toward possibilities that lie just on the edge of plausibility—the pig(oon)s don't fly, but they perform funerals and plan sneak attacks—Atwood's trilogy epitomizes genetic fiction's capacity to mold society in preparation for new scientific realities even as it insists on treating biology as an evolving expression of culture and its narratives.

The novels of Richard Powers examined in chapter 5 might seem to attract a completely different reading audience: they have moments of humor, but unlike with Atwood, one is never led to wonder, "How much of this is tongue-in-cheek?" Still, Powers's genetic metafiction shares several major characteristics with Atwood's, and it is little coincidence that she very favorably reviewed his National Book Award–winning *The Echo Maker* (2006). Like Atwood, Powers is profoundly committed to teasing out subtle cross-temporal relationships, with music serving as his constant dialogue partner both thematically and formally. And both novelists show how a more detailed picture of the cell might enlarge our culture's often impoverished, stultified concepts of the soul. Whereas Atwood is an unparalleled referee for exposing misapplications of scientific knowledge, Powers might be literature's most gifted athlete in reimagining science's positive potential. Powers's *Generosity: An Enhancement* (2009) and *Orfeo* (2014) challenge both over- and underestimations of new biotechnological possibilities by recontextualizing humanity in relation to much larger and smaller ecologies. The very structure of his fiction illuminates a process of bottom-up "compositing" by which the world makes new life and by which humanity makes new meaning. For Powers, these are deeply parallel processes in which can be found both freedom and preset inclinations; what his work offers is a vision of predisposed agency in which real character and profound conflict go together. One might say that the "creative nonfiction" undergirding his work is that the cell *is* the soul and the soul *is* the cell, constructions that refuse to let metaphysics consume knowledge and prohibit empiricism from denuding meaning. His characters and their metafictional journeys suggest how we might live by science and fiction simultaneously, embracing the subjectivity of language and interpretation while also making bold decisions founded on relative objectivity, sufficient warrant, and uncertain conviction. Indeed, such possibilities are unapologetically paradoxical.

By organizing, clarifying, and contextualizing a broad array of novels, films, and other works, this volume aims to shed indirect light on other past and future narratives that imagine biotechnological possibilities for remaking humanity and other species. *Editing the Soul* seeks to contribute especially to postsecular thinking and science fiction criticism, but its most basic goal is to draw greater, more informed attention to the rapidly mutating stories human beings are telling about their genomes. For that reason, this presentation of literary and cinematic criticism is a resolutely interdisciplinary project. In reaching beyond academic silos that produce conversations comprehensible only to relatively few, I am equally conscious that traversing boundaries like "literature and science" can too easily become mere assimilation of another field within one's own. Instead, by working from my own areas of relative expertise toward a two-way conversation with genetic biology and its applications, this book endeavors to illuminate both the science and the fiction constantly at work in biotechnology and its cultures.

1.
Genetics as Science, Ideology, and Fiction

Seconds after a boy's birth, a genome sequencing machine spits out odds that he will develop various medical problems, plus a shockingly precise projected life span. "Neurological condition, 60 percent probability; manic depression, 42 percent probability," a technician intones. "Attention deficit disorder, 89 percent probability; heart disorder, 99 percent probability; early fatal potential; life expectancy . . . 30.2 years." While vigorous crying and other solid Apgar scores might have represented full health according to past standards, this genome-based prognosis is devastating. The father changes his newborn's first name on the spot, waiting to pass on his own forename to later, biologically enhanced offspring. So begins the protagonist's journey in perhaps the most common fictional narrative used in twenty-first-century bioethics classrooms, *Gattaca* (1997). While this relatively low-budget science fiction film celebrates its hero's narrow escapes from genetic discrimination, in my experience, students tend to remember most vividly the overwhelming biopower of his dystopian future society. As the film's coda (cut from the theatrical release but available on the DVD) makes overwhelmingly evident, *Gattaca* is a warning that the eugenic mentality of the early twentieth century might return with even greater discriminatory force in the twenty-first.[1]

Gattaca's legacy reflects not only how effective falsehoods depend on some measure of truth but also how the deepest truths can be buried well beneath the facts. On one hand, the biotechnology projected by this premillennial tale is already proving accurate, as genome sequencing speed and precision continue to improve exponentially. Although cheap direct-to-consumer tests sold in recent years by companies like 23andMe sample around 0.02 percent of a person's genome, complete scanning of all of a person's six billion DNA bases is also moving rapidly toward widespread affordability. The first full human genome sequencing, part of the Human Genome Project, cost $3 billion and took thirteen years (1990–2003). As soon as 2007, though, Richard Powers's sequence would require only a few weeks and a price tag of only six figures. In 2010, Illumina began offering a machine capable of a full scan (repeated thirty times) for less than $10,000, and four years later, it reduced that fee tenfold, reaching the long-sought goal of a $1,000 genome test. Along with processing times now measured in hours or minutes, that cost is likely to keep falling.[2]

One part of *Gattaca*'s testing scene, then—the notion that genomic data could become so quickly and cheaply accessible that the average birthing center could provide it almost instantaneously—grows increasingly likely. Far more difficult, though, is the act of meaningfully interpreting the immense quantities of information thereby produced, and here both scientific and humanistic perspectives are essential. Niccol's film glosses over this part of the challenge, with the machine's immediate, purportedly unambiguous feedback reflecting an unqualified genetic determinism that the protagonist would spend the rest of his life resisting. Of course, as scientists learn more about the links between particular genes and various traits and diseases, the reliability of statistical predictions should continue to rise, but the scene exemplifies the dangers of perverting biology in an effort to dumb it down, whether for the 1997 audiences who were largely unfamiliar with such testing possibilities or for present ones who are often only marginally more informed. Put bluntly, beyond rare cases of early onset, single-gene terminal illnesses, the notion of producing such a hard, exact life expectancy for an individual is absurdly reductionist—whether instantly or after days of examination and whether today or decades from now. The enormous roles played by environmental influences and individual choices are almost completely overlooked in the society *Gattaca* imagines, and the fact that the protagonist rebels against the resulting injustice does not reduce the anxiety the film cultivates about biotechnology's apparent direction. Our hero is only the exception that proves the rule: we appear to be heading for a world where human futures are even more fully decided before birth.

If genetic fiction sometimes functions as a public relations mechanism for or against new genetic technologies, it's also its own means of uncovering knowledge and imagining new possibilities for transforming life. In 2004, Susan Squier argued, "We are now seeing a shift in the social valuation of science fiction, a shift in how we draw the line between 'fiction' and 'fact' that is related to the changing understanding of the human being produced by biomedicine. In short, the transformative processes of biomedicine are *enabled* somehow by the transformative narrative that is science fiction" (19, emphasis in original). More than a decade later, her approach is even more apposite. However one evaluates the biological accuracy or biocultural plausibility of films like *Gattaca*, the *feelings* they evoke remain relevant. Whether inhabiting the twenty-first century or a previous one, human beings have always worried that our destinies might be predetermined, that a mere glance at our bodies or our visible wealth might decide future opportunities. Thus the protagonist's evaluation for astronaut training, when he is told that his blood test *was* the interview, conveys the film's concern that new genetic knowledge coupled with social eugenics could strip humanity of individual agency, contra the film's tagline that "there is no gene for the human spirit."

A Brief Genetic Primer

Like *evolution* and many other scientific terms, *gene* does not mean the same thing today as it did two decades ago, nor is the term equivalent in the lab and on the street. The average nonbiologist's very basic definition might be, "A very small part of your cells that decides what traits you have and diseases you get." This is not entirely wrong, but it is not enough to imagine a single independent on-off switch deep within the cell, just awaiting easy discovery and manipulation. In fact, among scientists, a "gene" has become such a complex concept that some even question the word's usefulness. Here's a more accurate, intermediate-level definition: a gene is an intricate pattern of interdependent, often discontinuous DNA (made up of cytosine, guanine, adenine, and thymine) that can be used for building multiple proteins.

Notice that this formulation avoids confusing a gene with one of the individual C/G/A/T nucleotides that constitute it. Basic introductions sometimes suggest that genetic engineering is becoming a routine matter of flipping one of these bases to its opposite (say, an A to a T), thereby recoding a gene and in turn eliminating a terminal disease or changing a trait. Again, this is *not* completely misguided: relatively "simple" gene editing is rapidly becoming tenable, especially with the arrival of CRISPR as a delivery mechanism. Still, nonscientists rarely realize that most individual human genes are composed of *ten to fifteen thousand* base pairs, usually with many repeated sections. Adding to the complexity, these sprawling genes often overlap so that a single

C/G/A/T nucleotide can participate in multiple genes, and changing just one out of the ten to fifteen thousand can have cascading effects that exceed predictions. Thus it is all well and good to propose modifying a "SNP" (pronounced "snip" and standing for "single nucleotide polymorphism," one of only 0.1–0.2 percent of bases that actually vary between individual humans), but such an intervention could turn out to influence more than just the target gene.

There are many more intricate elements of intergene dependency, like epistasis, wherein one gene requires the presence of another in order to be expressed. The basic point, though, is that *a gene is not simply a single material button waiting to be pressed.* As Barry Barnes and John Dupré explain, "If genes are objects, then they are objects that vary enormously in their constitution, and they are *composite rather than unitary objects*—objects only in the way that the solar system is an object, or a forest is, or a cell culture" (53, emphasis added). They are composite objects because most genes function as interrelated patterns, groups, or sets, not as discrete material phenomena. Here's another analogy: nucleotides are like piano keys, and most genes are like piano chords rather than single notes. Instead of combining just a few keys, though, they normally involve ten thousand or more. And that's just the architecture of *one* gene: remember that a single human *cell* contains more than twenty thousand of these genetic chords, all being played in a composition that could sound different tomorrow than it does today. The upshot, as Barnes and Dupré point out via yet another analogy, is that

> there is no way that snipping a human genome into twenty thousand pieces will produce the twenty thousand genes it is now said to embody. To vary the metaphor, the set of genes in a genome is not to be compared to a bag of marbles. Take a marble from the bag and the rest will remain, but take a gene from the set and other genes could well prove to be missing as well. The DNA that by virtue of what it does is part of the gene for protein X may well be, by virtue of something else that it does, also part of the gene for protein Y. (55)

In short, there is a multiplier effect too easily forgotten in popular discussions of the genetic influences operating behind a given condition. When it comes to gene therapy, gene editing, and other forms of intervention, this means that there are both great opportunities and challenges ahead, with real risks of "curing" a disease but causing unexpected side effects.[3]

It is also worth cautioning that mastering the human genome is not just a matter of sufficient processing power. That barrier is gradually diminishing: bioinformatics has long since shown itself capable of utilizing enormous servers and complex algorithms to reveal promising statistical patterns, with genome-wide association studies (GWAS) highlighting SNPs of particular interest by comparing the genomes of people exhibiting a given condition with a control group. However, big-data research has raised as many perplexing questions as it has answered. Among the best known is the "c-value enigma," the fact that having a larger genome does not necessarily make one organism more complex than another. For example, there is a species of pine tree that would seem to be a far simpler organism than *Homo sapiens*, but its genome weighs in at forty-four billion bases, more than seven times the six billion of a human being. Nor is physical size the real factor in genome size, as there are also much smaller plants whose DNA quantities exceed ours. Indeed there has been a constant downward revision in the total estimate of genes comprising the human genome, from a hundred thousand a few decades ago to a little more than twenty thousand as of 2017. This is far from the tidy late twentieth-century picture in which one gene was imagined to equal one trait, and it means that scientists who succeed in untangling one strand of the relationship between a gene's transmission and its expression sometimes look up to find that other sections of the yarn have grown more twisted in the interim.

This is not at all reason to despair about the prospects for dramatic new medical interventions; rather, we should expect additional paradigm shifts ahead, not just upscaling of present knowledge. It is likely that more discoveries lie in the realm of so-called junk DNA, the non-protein-coding sequences that make up more than 90 percent of the human genome but were once screened out as peripheral noise. Relatedly, there is exceptional promise in the fields of epigenetics and proteomics, wherein greater attention is being paid to the processes whereby genes are registered, translated, and sometimes overridden by other cellular structures, bodily conditions, and external environmental influences. The role of the human microbiome is also a hot topic, given that the bacterial communities that share space with our bodies have total genomes several times larger than the human one and might exert considerable influence on human genetic expression. Some studies even question the longtime assumption that each of us possesses only a single genome; the only exceptions were thought to be rare chimeras, people whose earliest development involved one embryo absorbing another embryo with its own set of DNA, but scientists have begun investigating the possibility that genome mosaicism—variant genomes existing within a single person—is more common. However this plays out, the more one studies the rapidly transforming history of genetic surveying, testing, and manipulation, the clearer it becomes that the science remains in its childhood, if an extraordinarily precocious one.

Given the stage and rate of growth in biologists' knowledge, it is not surprising that popular representations are sometimes badly misshapen. That is to be expected, but the problem is that the loudest voices with the most stunning ideological claims often dominate public attention. When media outlets responded excitedly in the 1990s to Dean Hamer's attempts to isolate a "gay gene," for instance, one would have thought that doctors were only a few years from being able to instantly determine an embryo's eventual sexual orientation. In reality, even setting aside roles played by culture, birth order, and personal agency, the complex inclinations involved in sexuality are shaped not just by one nucleotide, or just one gene, but by many genes working together to produce a variety of contributing proteins. This is not to deny that homosexual (and heterosexual) desire is shaped by genetic inheritance—only to recognize that these relationships are more intricate than they are sometimes rendered. We should remember that genetic inheritance and environments have been coconstitutive forces throughout the history of life, so that even when "objective" discoveries have been announced, ideology has often driven projections of their significance. A 1997 item from the satirical news source *The Onion* still provides a crude but useful reminder here:

> BALTIMORE—On Monday, scientists at Johns Hopkins University isolated the gene which causes homosexuality in human males, promptly segregating it from normal, heterosexual genes. "I had suspected that gene was queer for a long time now. There was just something not quite right about it," said team leader Dr. Norbert Reynolds. "It's a good thing we isolated it; I wouldn't want that faggot-ass gene messing with the straight ones." Among the factors Reynolds cited as evidence of the gene's gayness: its pinkish hue; meticulously frilly perimeter; and faint but distinct, perfume-like odor.

The lesson in brief: there have been many "the-gene-for-X" announcements now, and generally the more complex the trait represented by "X," the more uncertain our determinations can be. It is also well worth asking who stands to benefit from such identifications.

Take, for example, Hamer's 2004 hyperbolically entitled *The God Gene: How Faith Is Hardwired into Our Genes*. While its cover promises hard biological answers to the question of why some people are believers and others are skeptics—a simplistic binary that ignores diversity within and across world religious traditions, changes in attitudes over an individual's lifetime, and many other nuances—the book's introduction admits the impossibility of the task. "The term 'God gene' is, in fact, a gross oversimplification of the theory," Hamer acknowledges. "There are probably many different genes involved, rather than just one. And environmental influences are just as important as genetics" (8). The irony is that the book's title and marketing nonetheless epitomize what Barnes and Dupré call "astrological genomics," a phenomenon whereby laboratory science bleeds into a simplistic biological determinism in popular culture. Just as the rigorous measurements and mathematics of astronomy are distinct from the cottage industry of crystal ball fortune-telling, genetic knowledge is too commonly confused with an inescapable fate. Deterministic approaches to genetics cheerfully gloss over realities like pleiotropy, the capacity of a single gene to contribute instructions toward the production of multiple traits, or the converse fact that most human characteristics are polygenic (influenced by multiple genes). While the simplistic rhetoric of genetic determinism might lead individuals to pay a direct-to-consumer testing company for an assessment of their genetic inheritance, people rarely receive quite what they expect. Sometimes there are intriguing results, and there are some single-gene disorders where a genetic test can be highly conclusive. However, most consumers garner little actionable data, at least presently, because of their lack of contextual experience with which to build an *interpretation*. For instance, the test might identify an APOE allele (or genetic variation) that has been linked statistically to Alzheimer's, but customers might not grasp that there are at least twenty other genes that have been similarly correlated, and the relative contribution of each is far from clear. Thus the test's indication that an individual has a twice-normal likelihood of manifesting the disease must be taken with a grain of salt, in the same way that (absent overwhelming exit polling data) an election should not be called with 1 percent of precincts reporting (especially when in this case, the "votes" of remaining citizens could change the impact of others already recorded). Even this analogy does not consider the impact often made by environmental influences and lifestyle choices.

Which brings me to a penultimate emphasis in this very brief genetic primer: even when there exists a strong correlation between a specific variation (allele) and a disease, in most cases, there will remain people whose genomes put them at high risk for manifesting the condition but never do. Single-gene disorders like muscular dystrophy, fragile X syndrome, and cystic fibrosis receive the most press in part because they are easiest to understand: in such cases, if a particular version of a gene is present, the individual almost certainly will manifest the disease. Even for these rare diseases, though, environmental influences and lifestyle choices can significantly delay or accelerate their appearance, not to mention their severity and speed of progression. Moreover, it is only relatively recently that whole genome sequencing, not just individual gene testing, has become a remotely viable consideration for the middle class, and it is changing assumptions about the meaning of data daily. Mendelian inheritance charts continue to have some use, but if classical physics can only account for a limited range of phenomena, the same is true of classical theories about dominant and recessive traits. Data from a dozen genes might combine to suggest a person has a 60 percent chance of developing depression, but she could also be voted "Sunniest Personality" by her senior class. The affirmation she receives from parents, siblings, teachers, and peers, not to mention decisions she makes about how to measure success, will play just as critical a role as her genetic inheritance, and scientists remain a very long way from being able to quantify that. In the genome age, we know more than ever before, but this expanded knowledge has opened just as many mysteries. The need for humility remains as urgent as ever, as does the reminder that being able to more accurately predict an individual's relative weaknesses will never justify withholding love or individual rights.

  This brief myth-busting exercise might have been relatively fascinating or unremarkable, depending on one's previous knowledge of genetics. Either way, it's worth noting that the more biologically informed a given artistic vision, the more heavily it tends to rely on readers and viewers to assess its scientific plausibility. That is, genetic fiction often utilizes devices such as irony and satire, but if a reader or viewer is unable to grasp the subtle distinction between what a novel or film is explicitly, technically saying and what it might mean between the lines, the work can fall flat or end up twisted beyond recognition. This is much like the challenge of reading a novel in translation: individual words and denotative references are easier to code and decode than less tangible, connotative devices such as voice or tone. In the same way that a story can depend on an audience to share a particular history and therefore recognize the import of a particular reference, works of art that engage new biological concepts sometimes expect readers to possess at least a passing familiarity with concepts of evolution and inheritance. At times, that knowledge even becomes a necessary foundation for fully understanding a plot, as with the evolving pandemic conjured by the CBC show *ReGenesis*. In other cases, one's biological background can shape subtler interpretive decisions, such as how we evaluate the appeal or danger of a pediatric cancer gene therapy trial. As a rule of thumb, though, the closer a text's temporal setting to the present, and the greater its realism, the more likely that a basic scientific knowledge will prove valuable. Genes rarely behave like the easy on-off switches for traits and diseases that our culture once imagined, and some sense of their composite, overlapping, often polyvalent nature is becoming immensely valuable for thoughtfully interpreting twenty-first-century biofiction.[4]

Beyond Determinism and Dismissivism

The quick primer in the previous section confronts some of the most common misconceptions about the science of genetics, but I would hasten to emphasize that we cannot understand the significance of genetics in isolation from culture. The meanings of genes and genomes have been negotiated across time and place, and it is misleading to imagine that there could be any biology without metaphors or narratives. The material world is there regardless of our names and measurements, of course, but our *understanding* of it is inevitably a subjective product of human language. As Squier emphasizes, "There is nothing inherently literary or scientific, only what disciplinarity makes so" (46). As a result, "every time an inside/outside divide is built between literature and science, we should study the two sides simultaneously" (46–47). In turning from the biological basics to some theoretical tools based in the humanities, my goal is not to assimilate one field within the other but to hold genetics and narrative together. To that end, the remainder of this chapter builds on the rapidly growing array of theoretical resources at the intersection of science and literature and then explains the book's reliance on postsecular theory.

To some, the perpetual "crisis of the humanities" might seem to make it more difficult to explain literary and film criticism's value to biology than the inverse. There is little question that global capitalism and the neoliberalization of higher education has raised the prestige of the sciences above that of the arts and humanities. Humanists have sometimes been our own worst enemies, having occasionally gone so far down the rabbit holes of high theory that we have lost touch with our colleagues in other disciplines, not to mention students and the general public. Nonetheless, scholars of culture and narrative have also made profound contributions to understanding the worlds cultivated by new technologies, and too often these have been overlooked. The intersections of biology and literature are particularly ripe for more such engagements, and in fact, they are badly needed. Whether one has personally undertaken some form of genome testing, seen a family member do so, or simply wondered about the choices of a movie star like Angelina Jolie—whose 2013 preventative double mastectomy was a much-publicized response to BRCA1 gene tests and a family history of breast cancer—today's stories about biotechnology have never been more widely compelling.

In assessing these narratives, especially those within literature and film, it is critical to shift from colloquial exaggerations about genetics toward a more holistic, culturally and personally contextualized vision. For most, this means questioning genetic determinism, but for others, this requires recognizing another danger, the less obvious but also pervasive force I call genetic dismissivism. Fatalism about genetic influence might be more common, but some sectors of Western culture persist in the equal and opposite habit of ignoring genetic influence. This might be explicitly related to the biblical literalism and cultural insularity that produces young-Earth creationism and climate change denial; it might be a side effect of the same fears that lead some to reject all childhood immunizations or other medications. Still, the problem is often less an active disdain for science or medicine than the result of a more innocent, often class-related lack of knowledge. Genetics is too often perceived as a high-tech science reserved for elite doctors or forensic detectives, not something that average citizens can or should understand at a basic level. The result is our culture's capacity to ridiculously exaggerate or trivialize the extent to which genes shape the patterns of human lives. Though perhaps less likely to cause immediate harm, ignoring genetic influence is as misguided as making a genetic profile an all-determining report on one's future.

Many of the best humanistic responses to genetics have come from bioethics. Dedicated to a philosophically and often theologically informed evaluation of the costs and benefits of particular biological discoveries and applications, this discipline has long bridged gaps among the sciences, social sciences, and humanities. It is no coincidence that federal budgets for genetic research have typically included a roughly 3 percent slice focused on ethical, legal, and social implications (ELSI), with bioethicists routinely winning these grants. While indebted to work in this area by such scholars as Donna Dickenson and Ronald M. Green, though, I ask different questions of the stories we tell about our genomes.[5] Beyond assessing the accuracy or advisability of biotechnologies as illustrated by literary and filmic works, I see these texts as inviting careful study in and of themselves. Novels and films actively *inspire* and *reimagine* science; they do not just reflect it retrospectively. For that reason, I am especially fascinated by the ways in which an expanding knowledge of genetics is reshaping narratives about the soul, its significance, and the nature of its relationships.

This is by no means completely uncharted terrain. Two of the literary critics on whose work I build heavily are Jay Clayton and Priscilla Wald. Fueled by a landmark NIH grant in 2003, their work on "genome time" in literature and its appeal to the "future perfect," respectively, has shown that however humanity engages new biotechnological possibilities, we cannot pretend to escape our past, nor is temporal influence truly unidirectional. Speculation about future forms of embodiment transforms humanity's present, as do new versions of history. The illusion of wiping clean the historical slate is the unaffordable dream of some forms of posthumanism, an expectation that humanity will reach a massive break when all that has gone before will disappear into a previous epoch. In such a vision, digitally and/or genetically enhanced descendants start fresh from a new, sterilized Eden, purportedly freed from the previous limitations of emotional unpredictability, subjective perspective, and physical endurance. There is a certain appeal to reaching escape velocity from our species' various atrocities, of course, but it remains elusive if not illusory. Genetic fiction takes a wide range of attitudes toward various forms of biomedicine and bioenhancement, but it is virtually unanimous in testifying that wherever human beings go, one way or another, our background narratives come along, even if only subconsciously or in the expectations created by others' memories.[6]

Moreover, while it is absolutely necessary for scientists to view their results with a degree of detachment, rigid objectivism is counterproductive for assessing those findings or devising new experimental goals. While scientific observers rightly seek to eliminate variables and attain the greatest degree of objectivity possible, the idealization of complete or absolute interpretive neutrality can be very misleading. Paradoxically to some, scientists' and entrepreneurs' successes in creating and harnessing new biotechnologies depend on embracing the uncertainties of existence and the unpredictability of inspiration, which also means resisting the logic of inevitability that assumes complete digitalized control should eliminate human caprice. To be clear, in questioning an unrelenting objectivism, my goal is not to water down the work of science, which depends heavily on rigorous control of samples and thoroughgoing statistical analysis. Rather, my aim is to maintain a firm grasp on the *metaphorical* significance of genetic biology, refusing to replace the overreactions of technophobia with an equally unwarranted surrender to technotranscendence. Briefly, there is just as much danger in an unthinking embrace of all biotechnology as in its unqualified rejection.

To synthesize humanists' insights about pre-1900 biological narratives, I lean on work by scholars like Markley, whose 1983 essay on Boyle and Newton, for instance, remains an illuminating piece about the Enlightenment's ideological impact. Against the supposition that this was the era in which reason overcame faith, Markley shows how science and theology actually emerged as "neither wholly complementary nor wholly contradictory" (356). More specifically, objectivism became science's undergirding belief system, "a necessary fiction that helps to structure [scientists'] perceptions of the physical universe" (357). The idea was that scientific philosophers should stand outside and above their experiments, manipulating only one factor at a time and relying on an entirely neutral gaze to record results. As a technical ideal, this was revolutionary. Markley shows, however, how Boyle and Newton also essentialized science as an end in itself, a means of reaching "an absolute and ahistorical knowledge" (364) that was appealing because of their "psychological cravings for certainty and stability" (366). If this begins to sound like religious dogmatism, that is no accident. What Markley sensed was how the rigorous processes of science, or *methodological* naturalism, gave birth to the ideological mistakes of modern scientism, a term I use to describe militant, exclusive forms of *metaphysical* naturalism. Science is foundational to Western culture, but scientism often rejects out of hand any knowledge claim or way of being that exceeds scientific verifiability. This distinction is foundational for my project: a genuine commitment to scientific rigor is very different from an ideological opposition to all forms of religion or a naïve technotranscendentalism, both of which can impede—ironically—the very social progress that new scientific knowledge and technologies can enable. Indeed, this clarification is particularly urgent the less familiar or tangible are a scientist's primary objects of study. When discussing genes or quarks, the slide into scientism tends to be even slipperier than when studying humpback whales or solar eclipses simply because of the greater abstractions and inaccessibility involved.

A core problem for the public rhetoric of genetics, then, has been its overreliance on transcendent language. Perhaps we should not be surprised: Arthur C. Clarke noted in 1961 that for nonexperts, any sufficiently advanced technology will prove indistinguishable from magic. Still, well before Hamer popularized the "God gene," the extent to which simplistic technoutopian ideals were informing visions of genetic engineering was already remarkable. No one has pursued this historical pattern more doggedly than José van Dijck, who traced the Cold War–era emergence of genetic engineering as a kind of scientific antidote to the threats created by nuclear warfare. Writing near the turn of the millennium, van Dijck recognized the emergence of a pseudoreligion in the "metaphysical references" and "abundant use of religious imagery" in James Watson's autobiography about the discovery of DNA's double helical structure. Regarding Francis Crick's choice of the term "central dogma" to describe the rules by which DNA passes on genetic information, she explained that "the very idea of a set of divine laws enabled the concept of scientists reading God's will writ in nature, like priests reading his will in the Bible" (48). Furthermore, she reflected, "Like Christianity, the Central Dogma has its prophets, its messengers and its message—every ingredient for a secular belief," and so "while Catholic and Protestant theologians attune their religious standpoints to the new genetic principles, geneticists borrow the language and imagery of religion to present themselves as 'helpers of God'" (49). Van Dijck was among the first to sense technoculture's divine pretensions shifting from nuclear physics to genetics, and her work offers a comprehensive narrative of that transition.

In some ways, this is simply a reflection of a larger twentieth-century shift that is increasingly clear to historians of science. David F. Noble, for instance, explains that "modern technology and modern faith are neither complements nor opposites, nor do they represent succeeding stages of human development. They are merged, and always have been, the technological enterprise being, at the same time, an essentially religious endeavor" (4–5). Noble sounds much like Squier insisting on the coconstitutive nature of science and literature, or for that matter, Haraway provoking more integrated thinking about "natureculture." The surprise for some is in bringing religion into the conversation. Such metaphysical framings of genome-age culture could seem an overstatement, but for historians, there is little question that biology's achievements have long utilized religious flourishes in their public self-presentation.

None of these scholars is claiming that humanity has literally built or should build a set of traditional religious doctrines around the discoveries and rhetoric of biotech. Rather, this field has often *functioned* as the locus of an alternate theology. To hear direct-to-consumer testing companies tell it, your genome is the ground of your being, the very center of what makes you a unique person. Coupled with a rigid enough social system, your genes might even decide, as the 2003 film *Code 46* imagines, whom you should and should not love. In a trailblazing 2000 article, Wald adeptly balanced the laboratory science with such affectations in its public representation: "Like all creation myths, genetics (again, broadly conceived) . . . re-poses traditionally religious or philosophical questions of will and determinism in the language of science. . . . To understand genetics in this way is not to downplay the importance of the medical technologies it makes possible. Rather, it shows how that importance is compounded by the importance of the representational technologies through which it registers and partly transforms world views and through which it risks becoming a belief system in its own right" (705). Wald's insight, like van Dijck's, is now even more difficult to contest: genetic images and metaphors perform cultural work that heavily overlaps with the labors of traditional religions. This does not mean the science is any less legitimate, but it suggests why questions that once seemed the separate domains of biology, religious studies, and literary criticism now demand simultaneous attention from all three (not to mention history, philosophy, and other fields).

What keeps these metaphysical visions of the genome going? There is a deep irony here. Several of the novels analyzed in this book reveal that one of the most enduring sources of genetics' religious connotations has been the antireligious rhetoric of some of genetics' leading figures, from Watson and Crick to Craig Venter and Richard Dawkins. By equating the work of science with the militant ideology of the New Atheists, such figures have often asked biological metaphors to fill the cultural vacuum created by evicting religion. This says nothing about their significance as scientists: Watson and Crick's discovery of the double helix is legendary (as should be the work of their colleague, Rosalind Franklin). Venter is rightly famous as the private researcher whose company's technical innovations pushed the publicly funded Human Genome Project to a far earlier completion than it would have otherwise achieved, and he has gained more recent attention for promising efforts to develop clean energy via synthetic biology. Meanwhile, Dawkins is well known for his groundbreaking 1976 book *The Selfish Gene*, but his scientific work has sometimes been preempted by his culture warrior alter ego and his attacks on all things religious, with minimal effort to distinguish between the vastly different forms that fall under that banner. How did the gene become a synecdoche for the soul? It involved some of our most brilliant scientists twisting the scientific method into an all-sufficient ideology and reacting to fundamentalist language about warfare and domination with equally absolutist rhetoric.

For Venter, the enduring enemy has been the problem of "vitalism."[7] In *Life at the Speed of Light*, he rightly insists that science must follow wherever the data leads, but he overgeneralizes in apparently questioning the naturalistic commitment of *all* religion. Starting with nineteenth-century chemist Friedrich Wöhler's *experimentum crucis* as a landmark in defeating vitalism that foreshadowed his own development of a synthetic organism, Venter celebrates any effort to show that the techniques of biology and chemistry are fully reliable and sufficient routes to understanding the processes of life. In this sense, he might have been a valuable witness at the 2005 intelligent design trial in Pennsylvania, where his colleague in biology Kenneth R. Miller, a committed Catholic, helped a George W. Bush–appointed judge grasp how intelligent design's rejection of evolution was ultimately creationism in another guise. In fact, if in his obsession with defeating "vitalism" Venter meant only refusing to replace available scientific data with metaphysical speculation, I would be fully behind his effort. Science education cannot do without biology's central paradigm. However, the problem is that by "vitalism," Venter apparently means any and all reference to that which is not materially testable. Rejecting the act of belief wholesale rather than simply rejecting beliefs that are demonstrably incorrect, such rhetorical moves make unnecessary enemies of religious and secular forces that could together accomplish a great deal of good.

This is where the more nuanced epistemology featured in many genetic fictions and articulated by postsecular theory becomes so attractive. In the twenty-first century, it is insufficient to pretend with Enlightenment objectivism that knowledge can be acquired independently of all faith or intuition, that meaning can be entirely separated from the particulars of one's interpretive position. The scientific method has proven invaluable for bracketing out the whims of subjectivity, providing us with what I call "relative objectivity."[8] Of course, objectivity remains a methodological ideal: good scientists run the experiment over and over, reconfiguring constantly so as to account for as many external variables as possible. But as we will see repeatedly in the chapters ahead, good scientists do something more than this when they consider the significance of their work for society and ask what projects they should attempt to get funded next. In such contexts, despite all the double-blind trials in the world, they remain human beings with very specific, context-dependent goals and desires, and they inevitably participate in humanity's metaphysical efforts to unite far-flung areas of knowledge.

My postsecular approach to epistemology is that a certain form of "belief" is not science's antithesis, nor a sufficient replacement for science, but science's *prerequisite*. There are no hypotheses without hunches that nonobvious things might be true. Without the ability to take risks in pursuing uncertain hypotheses—where "faith" means not an antonym for knowledge but an enabler of it—one will never achieve meaningful discoveries. From my postsecular angle, the best science and the best religion are *action verbs* rather than static nouns, and they are driven by curiosity and intuition rather than imprisoned by fear or dogma. This clarification is so pivotal for what follows and could be so easily misunderstood that even before we reach the book's fullest exposition of postsecular theory, I will pause to stress even further what it does *not* mean. By arguing for postsecular definitions of faith and knowledge and by aligning belief with the intuitive leaps that undergird scientific discovery, I do not wish in any way to question the idea that there are such things as "facts." This book does not attack the scientific method any more than it invites a "God of the gaps" approach, whereby religion steps in wherever scientific mysteries remain. It does not elevate the human practice of religion above the human practice of science—or vice versa. Most of all, I would emphasize that my aim is not to broker some sort of superficial, wishy-washy compromise between science and religion in the interest of better public relations. I'm proposing something far more specific—namely, challenging articulations of science and religion that unnecessarily misrepresent or devalue *either one*. The innumerable human activities that fall under the banners of "science" and "religion" have real differences in purpose and methodology, and those must be respected. At the same time, the postsecular notion I advance here is that while scientific and religious viewpoints are not always relevant to one another and should never be subject to one another, they are not necessarily incompatible. The assumption that they must be at war is in fact a major part of the problem, a mistake that is very particular to relatively recent history, especially within the United States.

I also want to begin suggesting here that once we step back from the empirical work of a field like genetics and start interpreting the data's significance for individual human futures, we are doing something more closely akin to interpreting literature than it might appear. Both processes benefit enormously from careful consideration of one's epistemological assumptions. By contrast, in rejecting all forms of faith as inherently irrational, Venter equates religion with antiscientific thinking, as in formulas like "When there is mystery, there is an opportunity for vitalism and religion to thrive" (130). As is evident here, the long-standing secularist tradition of lumping an enormous range of cultural expressions into a single phenomenon does far more damage than good. Dawkins has also relied on this move to reduce all religiosity to a single phenomenon, as when he linked arms with Daniel Dennett, Sam Harris, and Christopher Hitchens in the mid-2000s. Calling themselves "the Four Horsemen," they penned a set of popular post-9/11 manifestos damning not just religiously inflected terrorism but any effort to find meaning that reaches beyond provability. A major weakness these volumes often exhibit is inattention to the ideology inherent to all linguistic choices, the impossibility of converting every shade of meaning into binary code. One of Slavoj Žižek's most useful insights is that ideology is operating most powerfully just when you think you have escaped it. In Venter's case, the overstated rejection of all religion makes it difficult not to hear his own form of messianism operating. How else to read descriptions—and these span only a few pages—of "the course of the history of genomics" (50) changing because of a chance encounter he had at a 1993 scientific meeting; of an audience that "rose in unison and gave me a long and sincere ovation" at a 1995 conference, prompting his reflection that "I had never before seen so big and spontaneous a reaction at a scientific meeting" (52); or of a 1996 genome study by his lab that "appeared on the front page of every major paper in America and made headlines in much of the rest of the world" (56)? Venter's genius deserves genuine appreciation, but when science becomes scientism and is coupled with a seeming desperation for adulation, it is little wonder that many find the mixture dangerous.

At one level, there is an enormous chasm between contemporary research's steel-and-glass corridors and the bloody altars found in many religions' histories. Their slaughtered animals and today's massive server arrays seem centuries apart, but there are more epistemological similarities than many wish to admit. Too few are Haraway's modest witnesses, characters like the psychiatrist in Terry Gilliam's *Twelve Monkeys* (1995), who frankly acknowledges that her discipline can inflate its authority to cultic proportions: "What we say is the truth is what everybody accepts. Right, Owen? I mean, psychiatry: it's the latest religion. We decide what's right and wrong. We decide who's crazy or not. I'm in trouble here. I'm losing my faith." If scientists do not own up to uncertainty, when they do not embrace it as a badge of honor rather than evidence of incompetence, it falls to others to point out how science's shift from methodology to ideology ends up disembodying the self. Venter is wonderfully compelling when he imagines using digitalized biology to study alien life-forms isolated by robotic probes on other planets, but we must question assumptions that nothing real can escape our measuring tools. Likewise, Dawkins is a wonderfully creative biologist, a staunch defender of his discipline's achievements, but as an antireligious prophet, he risks confirming the worst fears of antiscientific extremists, driving them further into their rebellions against public education and the overwhelming consensuses of evolutionary biologists and climatologists.

Van Dijck is particularly evocative in recounting the precise means by which genetics has sometimes become lost in the thickets of its information technologies. During the Human Genome Project and its immediate aftermath, she explains, "The gene metaphorized into the 'genome', genetics into 'genomics', and the geneticist became an amalgam of a molecular biologist and a computer scientist." This meant that "whereas 'information', in the 1960s, had served as a metaphor, it now became material inscription," and "now that the computer dominated the genetic imagination, the body became part of an informational network." Most recently, the consequence has been that the map trope has been "used so often that its figurative meaning escapes the reader's notice and reaches the stage of demetaphorization." Scientists and nonscientists alike lose any sense of narration, story, and approximation and instead begin talking about "'sequencing a region', 'guideposts', 'disease loci' and 'genetic markers' as if there are actual sites to which illnesses can be reduced" (120–21). Also overlooked is that many of the processes involved still remain beyond current understanding—that in fact, the core of individual identity might be too dynamic for binary code or conventional computing to represent. In the near term, the digital conceit of genetics lets scientists "tacitly change the primary goal of the mapping project from defining the ideal (healthy) human being into defining its diseases or flaws," which ultimately "yields a view of the body as the flawed version of the perfect code, and concurrently holds the promise of an easy genetic fix" (122). Biology thereby takes on a deeply evaluative and in fact condemnatory mode, with medicine becoming a source of twenty-first-century elixirs and doctors comprising an involuntary priesthood holding the keys of the kingdom rather than a group of fellow human beings whose knowledge and capacities are tremendous but still limited.

If van Dijck's historical labors question Venter's ideological tendencies, N. Katherine Hayles offers an even more incisive critique of Dawkins's vision. Van Dijck had already shown that "Dawkins's introduction of the selfish gene does more than simply revamp the root metaphor of the factory [that was already driving genetic research]: it also inserts business and management idioms into what is still basically perceived as an object of scientific research" (93). Hayles's contribution is not just to note additional problems with commercializing the genome (a train that left the station long ago) but also to point to an alternative epistemology that reestablishes the value of uncertainty. Challenging Dawkins's representation of metaphor in *The Selfish Gene*, she says it reflects a condescending "giftwrap model of language" wherein "I wrap an idea in language, hand it to you, you unwrap it and take out the idea" (147). By contrast, Hayles insists on viewing narrative tropes as the unpredictable, dynamic bonds between our thinking minds and our physical selves, our spirituality and our materiality: "Metaphor is not opposed to scientific work but intrinsic to it. Metaphor performs essential functions in orienting and guiding thought; *it connects abstraction and embodiment*; it allows us to discover regularities between what we perceive and what exists outside of ourselves; and it entwines cultural presuppositions with scientific frameworks. These complex functions can be summed up by saying that metaphor works to connect and contextualize, broadening the space of abstract thought by embedding it in physical, sensory, linguistic and cultural contexts" (144, emphasis in original).

For Hayles, evaluating the ever-shifting meanings of metaphors follows naturally from a commitment to carefully measuring the material world. Her approach sets the stage for this book's turn to postsecular thinking by recognizing the possibility of the physical and the metaphysical inhabiting and underscoring one another, without one's consuming the other. Hayles resists Dawkins's semignostic effort to "identify the gene as an actor distinct from the individual, who is re-conceptualized as a remote-control mechanism operated by the gene" (148). Instead, she puts greater stock in individual agency, the human creativity capable of making inexact but illuminating comparisons. Rather than viewing the measurable and the immeasurable as inherently contradictory, Hayles exhorts us to keep thinking about what knowledge of our genetic patterns can and cannot provide. And she maintains her balance on this tightrope by holding together "metaphors" with "constraints." Her idea in labeling this epistemology "constrained constructivism" is that "reality is never present to us as such; rather, our sense perceptions are self-organizing processes that construct the world we know from the unmediated flux, unknowable in itself." Echoing Markley's critique of the ideology of pure objectivism, she explains that although we must constantly revise our scientific models, "we can never know if these models are identical with reality, because we cannot occupy a position from which we could encounter reality independent of our perceptions. Rather, the best we can do is determine if our models are *consistent with* the unmediated flux as we experience it, a proposition that indexes our observations to the range over which we observe phenomena, the nature of our sensory and perceptual apparatus, the languages available to us, and so forth" (157, emphasis in original). Human beings are using genetic knowledge even as we radically revise it, and in that process, we flit quickly between positions that are in and out of control. It is crucial that we combat efforts to freeze scientific knowledge in place and let it harden into an unquestionable, all-encompassing ideology.

In sum, leading scholars of science and literature such as Markley, Wald, Squier, Haraway, and Hayles have been cautioning that, as van Dijck puts it, "the metaphor of DNA as language, of the genome as a Book of Life, promises the possibility of a final solution to the uncertainties of living" (152).[9] Far from condemning genetic research or applications, they encourage the pursuit of genetic answers that shed light on mystery without presuming to eliminate it. This should not seem a radical proposal, but it is all too easy to forget, not just in the abstract realms of criticism and theory, but also in the most basic areas of life. Michael Bérubé's scholarly autobiographies illustrate these stakes especially poignantly. Writing about his son Jamie, who lives with Down syndrome, Bérubé was already arguing in the mid-1990s against the overreaching predictions of a eugenic mentality like the one soon envisioned by *Gattaca*. "I do not want to see a world," he confessed then, "in which human life is judged by the kind of cost-benefit analysis that weeds out those least likely to attain self-sufficiency and to provide adequate 'returns' on social investments" (52). Rather than surrender to the endless abstractions of utilitarian thinking, Bérubé provided one more example of how the meaning of genetics depends on historical specificity and physical embodiment. We need to probe uncertainty, not destroy it: it is too valuable for cultivating curiosity and opening new avenues for both scientific and humanistic discovery.

The Postsecular Payoff

At its core, postsecular theory is an effort to reinvigorate discussions of contemporary belief and nonbelief by recovering the positive value of uncertainty.

One prerequisite, of course, is the willingness to grant that Western culture is neither as simplistically religious nor as secular as it is often portrayed. In fact, debates between "religious" and "secular" positions, including those between theological and biological viewpoints, inherently involve historically and geographically specific constructions. As Michael Kaufmann explains in one of postsecular theory's clearest early manifestos, "There is no idea, person, experience, text, institution, or historical period that could be categorized as essentially, inherently, or exclusively secular or religious," and as a result, "what counts as 'religious' at one time and place may count as 'secular' in another" (608). By extension, I am suggesting, the lines drawn between religion and science are also the results of specific cultural negotiations.

That might or might not seem a shocking claim; in academia and especially my discipline of literary studies, though, it is far less common than awareness of the cultural construction of gender or race. Unlike English professors' routine resistance to easy oppositions of "masculine versus feminine" or "whites versus blacks," binaries like "sacred versus profane," "holy versus ordinary," and "spiritual versus physical" are often accepted matter-of-factly, with one side rendered the normative default and the other an alien deviation. In American culture more broadly, such moves *seem* useful for both sides of rhetorical battles, not least because they silently exclude third, complicating possibilities. But just as categories like "queer" and "transgender" trouble facile treatments of sexuality, we need to make room for terms like "atheistic Jew" and "theistic evolutionist" to enrich understanding of both religion and science. In fact, failing to recognize the dynamism of this spectrum is very costly. Historians have detailed how the early twentieth-century United States became fixated on battles between fundamentalism and modernism, cultivating the bombastic spectacle and self-contradictions of the Scopes trial as well as other regrettable elements of 1920s culture like Prohibition, but it is sometimes easy to overlook the broader social consequences of religious illiteracy and epistemological shallowness in our own time. Religious-secular rhetorical warfare today continues to reflect and exacerbate the ideology of objectivism, relying on the illusion of the completely neutral, unmoving, unaffected interpreter. This book is not about the resulting political landscape; suffice it to say that I see ties between oversimplification of religion-science relationships and the level of populist rage and conservative-liberal polarization that presently characterizes American culture. By contrast, postsecular thinking rejects the fallacy of the voice from nowhere, whether that position is claimed by theist or atheist. It illuminates the particularity of specific religions and secularisms and points out both the unique blind spots and the insights afforded by specific backgrounds of experience. And as a result—to come back around to my real focus—it can expose how appeals to technotranscendence around genetics and its fictions simultaneously challenge and rely on religious myths.

Earlier I noted that science has been in conversation with religious thought since the beginnings of the Enlightenment, but that tension has only grown as bioengineering has come closer to rewriting the very fiber of human bodies. If the poetry of Psalm 139 is taken to promise that God literally and materially "formed my inward parts" and "knit me together in my mother's womb," Jewish and Christian theologies might seem necessarily antagonistic to biotech's growing capacity to inspect, select among, and manipulate embryos. However, this reaction reflects the character of the specific religious moment within which these "secular" technologies have emerged. Several centuries earlier—I am thinking of a tradition running roughly from Francis Bacon to Joseph Priestley—more theologians from these traditions would have been likely to regard the capacities of contemporary bioengineers as divine gifts. Such counterintuitive responses continue today, but they are less visible in our media. Most often, references to "religion and science" bring to mind figures like Ken Ham and Bill Nye debating the age of the Earth at Cincinnati's Creation Museum, Bill Maher making fun of evangelicals in *Religulous* (2008), or Sarah Palin rejecting climate change science. What such scenes obscure is the extent to which true knowledge precludes treating either scientific or theological statements in a void, absent of attention to genre or historical context. Just as it is enormously problematic to take the poetry of the Psalms—not to mention the two creation myths at the beginning of Genesis—as if they were attempting to contradict modern astrophysics, geology, or biology, the interpretive move that necessarily equates genetic testing or modification with usurping God creates far more problems than it solves. Instead, we must accept that interpretive choices are built into the work of science and theology alike and that neither truly benefits from performing an impossible certainty.

Postsecular theory's value to understanding science's significance, then, lies in demonstrating that the metaphysical rhetoric sometimes adopted by biotechnologists and more often by their proponents in popular culture is a direct result of how a culture delimits "religion" in relation to other aspects of life. If we assume the purview of religion to lie only within the institutional limits of church, synagogue, temple, or mosque, we fail to grasp its influence elsewhere. Rather than just defining religion substantively in terms of official doctrines and practices, it is especially critical in the twenty-first century to discern its functional presence in less obvious contexts. As Graham Ward observes, "Religion does not live in and of itself any more—it lives in commercial business, gothic and sci-fi fantasy, in health clubs, themed bars and architectural design, among happy-hour drinkers, tattooists, ecologists and cyberpunks. Religion has become a special effect, inseparably bound to an entertainment value" (132–33). Ward gestures to the uncanny parallels between the mystical technologies of the cinematic megaplex and the sublime lighting and sound systems of the contemporary megachurch and then asks, What's the difference? Rather than dismissing all religion as an addiction of the weak-minded, he invites its reconsideration as a broader set of human activities. As with Paul Tillich's redefinition of religion as matters of "ultimate concern," the effect is to scramble easy divisions between the religious and the secular and to expose these categories' coconstitutive nature. In the late-career words of Jacques Derrida, Western culture cannot continue to "oppose so naïvely Reason *and* Religion, Critique or Science *and* Religion, technoscientific Modernity *and* Religion" (65, emphasis in original).[10] Too much depends on discovering and then acting on the common ground regularly shared across religious (and nonreligious) differences—not as motivated by a superficial political correctness or in service to an amorphous civil religion but as a full-throated expression of the true sacredness of individual persons and communities.

This chapter began with the birth scene in *Gattaca*, where we saw how the potential gifts of genetic testing might be co-opted by the eugenics of a corporate state. With postsecular theory in mind, let's return to that scene, looking around its edges to see how much more becomes legible. Tellingly, the moment is introduced by a God's-eye view of the protagonist's conception, with Vincent-cum-Jerome marveling in the voice-over, "I'll never understand what possessed my mother to put her faith in God's hands rather than those of her local geneticist." It is not just that there is a religious flavor to the scene preceding the baby's birth and genetic testing; the cinematography and editing is more nuanced than that. The visual iconography begins with a couple's loving gaze in the back of their Buick Riviera and then fades into a rosary with crucifix, seeming to bless their traditional conception (fig. 4). From there, we shift to the sterility of the birthing room and its instant, unambiguous evaluation of a child's future. Before he can be placed at the breast, he must have his heel pricked, a shot in which the actors are blocked so that his mother is eclipsed

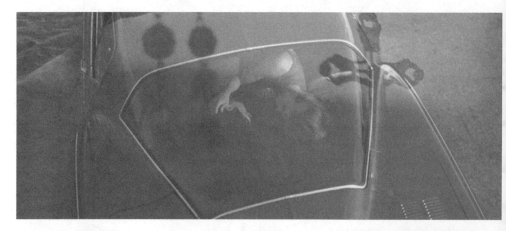

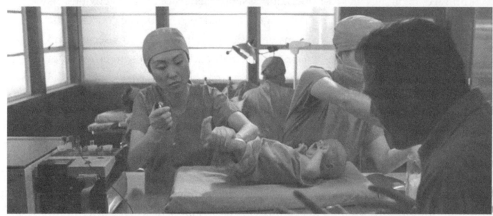

Figs. 4–5 *Gattaca* (1997) juxtaposes the intimacy of conception against the separation of the birth center and its genetic testing protocol.

by technicians and only his father's silhouette looks on, both meeting his son for the first time and awaiting his statistical assessment (fig. 5). While the glass that separates him from wife and child is not obvious, he is held at a distance, and in some frames, there is a subtle glare. He is able to hear and be heard, but the effect is to align the scene not with the audience's future but with a past America in which husbands were excluded from birthing rooms. Despite the depicted biotechnological advances, the film intimates that this is a cultural step *backward*.

But even as *Gattaca* calls attention to patriarchy within families and the medical establishment, religion is not simply equated with nor pitted against authoritarianism. On one hand, the violence referenced by the crucifix might be understood as the ultimate form of child abandonment, making the father's last-minute hesitation to pass along his name another critique of a calculating, paternalistic religion. On the other hand, we could hear the protagonist testifying that his mother chose God over genetics, and we would then conclude that *Gattaca* is relieving the cold measurements of paternalistic biotechnology with a warmer, more authentic maternal religiosity. What both of these interpretations neglect, however, is the precise rendering of the line in question. Vincent-Jerome does not say that his mother chose faith over reason but that the *object* of her faith was atypical. In her choice of "God's hands rather than those of her local geneticist," the question was not whether to have faith but in whom to place it. Our knowledge is limited, the film acknowledges, and sometimes we must take leaps into the unknown. Sometimes we are convinced that we have chosen well; sometimes less so. Either way, we can't be sure. The roads diverge, and we can only try one.

In other words, I wish to set aside both the kind of traditional religious reading that might view *Gattaca* as placing ethics above scientific measurement and the secularist one that would treat the film as a testimony to religion's patriarchal abuses. Instead, I think a more insightful postsecular reading is that the film illuminates *both* the distinction between faith and knowledge *and* their interdependence. The payoff of this approach becomes particularly evident when one researches the film's compositional process. There have been many book chapters and articles about *Gattaca*, but to my knowledge, none examines the version of Niccol's screenplay available online, which reveals fascinating differences in the narration about where one should place trust. Contra the film's actual voice-over, the screenplay reads, "Those were early days—days when a priest could still persuade someone to put their faith in God's hands rather than those of the local geneticist." The temporal distinction in this version of the line reflects a common iteration of secularization theory—common at least before the election of George W. Bush and the events of 9/11—that religious influence in America was dissipating and that it would eventually prove a relic of more primitive civilizations. The film's final version wisely took a more subtle route, but the screenplay's repeated associations of religion with technologically enforced normativity are illuminating. Vincent-Jerome also laments his "genetic prophecy"; when his parents visit their genetic counselor as part of the in vitro fertilization (IVF) process, the business is named "PRO-CREATION"; discussing his learning disabilities, our hero recounts a "genetic scarlet letter" following him from school to school. Furthermore, the film's voice-over was to include an explanation that while nongenetically enhanced children in Vincent-Jerome's society are officially called "invalids," they are colloquially known as "godchildren," "men-of-god," "faith births," "blackjack births," "deficients," "defectives," "genojunk," "ge-gnomes," and "the fucked-up people." Clearly, Niccol once imagined a far more pronounced antireligiosity operating in his dystopian society's judgments of inferior bodies. A postsecular approach to the film does not use this observation to argue that the film is "really religious" (in its critique of this dystopian society's impiety) or "really secular" (in demonstrating the ineffectual results of "God's hands rather than those of the local geneticist"). Instead, it focuses attention on the relationship between the religious and the secular/scientific, asking how this futuristic film portrays that tension and what that negotiation can tell us about our circumstances today.

One more element of *Gattaca*'s screenplay especially rewards a postsecular approach. The film's final cut never quite makes clear why it matters for Vincent-Jerome to become an astronaut. The easy answer, of course, is that this profession represents the upper limits of a meritocracy, that the character might just as easily have clawed and scraped to become president or perhaps an eminent concert pianist like the one whose six-fingered glove his girlfriend Irene catches. However, the screenplay indicates a more significant connection between its protagonist's vocation and the film's genetic concerns. At one point, the character whose identity Vincent-Jerome purchased remarks, "I still can't believe they're sending you to the Belt—you of all people—never meant to be born, on a mission to discover the origin of life." This cryptic reference to the mission's purposes grows clearer later in the screenplay when Vincent-Jerome visits the spacecraft he will soon launch. The director of spaceflight picks that moment to wax poetic about the mission's ultimate significance: "Somewhere in the dust of Gaspra is the key. Back to the beginning of the book—the life we became. With the original building blocks, who knows how far we can take 'the godding.'" What this unfilmed scene might have emphasized (though with a heavy dose of melodrama) is that *Gattaca*'s dystopian society does not just determine individual opportunity entirely on the basis of superior genes, which are themselves tied directly to a family's wealth. The film also makes a curious, perhaps unconscious twist on James Watson's famous remark, "We used to think our fate was in our stars. Now we know, in large measure, our fate is in our genes." *Gattaca* would have invited us to reverse Watson's logic, suggesting that since genes do *not* end up revealing the protagonist's fate, it really *does* depend on the stars. We will travel into space in order to learn of ourselves what our genes cannot tell us: who we are, where we came from, and why it all matters. In this sense, coupling *Gattaca*'s screenplay with the final film both embraces and complicates its push toward technotranscendence. The film acknowledges the likelihood of a deterministic obsession with genetic influence, but it seeks to overshadow that danger with an appeal to heroic determination and a pursuit of humanity's ultimate origins.

This book's driving impulse, then, is that genetic fiction can be an invaluable tool for accessing and influencing our culture's biotechnological imagination. Particularly when interpreted with an eye to questions of ultimate concern, it reveals how the meaning of genetic inheritance is always culturally mediated, even if it is often messy and confusing. Stories that take genetics seriously are beginning to shift beyond linear narratives that anticipate a completely separate posthumanity as well as beyond merely circular visions that dread an endless recurrence of the same. Their shapes are turning spiral, so that they integrate simultaneity and sequence and evoke repetition with difference. Such structures convey incremental progress and regress as well as bigger lurches forward and backward. Even as they acknowledge the rise in our capacity to anticipate the future, they constantly challenge assumptions of inevitability. The cumulative effect is a deepened sense that every story must remain evolutionary, ongoing, *living* rather than nailed down, incapable of mutation, *dead*. Moving beyond genetics as an objectivist ideology means learning how to live by myth but not in denial of facts. Of course, no genetic fiction is an infallible guide in and of itself. Still, rather than being taken unawares by new biotechnologies and merely reacting in the moment, we should pay close attention to works that can illuminate their potential blessings and curses. In the process, we are likely to learn how to live more wisely now, regardless of which biotechnologies we choose to employ or decline.

2.
The Evolution of Genetic Fantasy

If *Gattaca* aligns religious and scientific language by comparing God's hands to those of a geneticist, the unwritten rule of stories about human clones is that they and their creators must have biblical names. Mulder and Scully chase down clones named Eve in a 1993 *X-Files* episode, Arnold Schwarzenegger plays multiple Adam clones in *The 6th Day* (2000), and Robert De Niro's scientist character names his clone boy Adam in *Godsend* (2004). The "father" in the latter film, of course, had to be Paul, another name used frequently in clone narratives, from Katherine McLean's early short story "The Diploids—Die, Freak" (1953) to the recent television series *Orphan Black*. Indeed, the latter BBC America hit, to which I return in chapter 3, is a veritable roster of biblical names, variations of which appeared in many earlier cloning tales as well. Among others, these include Sarah (also used in *The Island*, 2005); Rachel (also Rachael in *Do Androids Dream of Electric Sheep?*, 1968, and its film adaptation, *Blade Runner*, 1982); Mark (*Where Late the Sweet Birds Sang*, 1976); and Marian (also Miryam in *Solution Three*, 1975, and Maria in *The House of the Scorpion*, 2002, and *In His Image*, 2003).

The list could go on, but why are these biblical namesakes so pervasive in fiction about biotechnology? Even this brief sample should suggest more than coincidence: no matter how disinterested in traditional forms of religion authors and screenwriters consider themselves, they are routinely compelled to endow acts of human genetic "duplication" with a sacred aura. Once upon a time, critics inspired by Northrop Frye might have taken such a pattern as evidence that the Bible remains the "great code" of Western literature and popular culture; more recently, others might be prone to argue that the trend evidences technoculture's eagerness to take over the task of feeding the masses, or at least their imaginations. Alternatively, one might read these convergences of scientific speculation and religious vocabulary as ironically playful, even sardonic. In any case, they trouble the easy religious-secular binaries to which we are accustomed, acknowledging that metaphors and archetypes associated with religion have deeply permeated other cultural spaces and that there can be no secularism without specific (even if unstated) forms of religiosity against which to define itself.

Rather than rehashing the tired debates between the religious and the secular—as if it were possible to prove the ascendance or decline of such a global, dynamic, polyvalent category as "religion"—this chapter charts a path through genetic fantasy, the subgroup of genetic fiction that dominated the twentieth century and still shapes many thriller novels and blockbuster films. Within that sizable field, we focus on the evolution of stories about reproductive human cloning, the prominence of which might be tied to that practice's relatively unanimous repudiation by bioethicists. It is not hard to find a wide range of for and against positions about stem cell research's use (or indefinite storage) of early-stage embryos, and there are many insightful debates about various forms of genetic modification. Historically, however, the idea of human reproduction via somatic cell nuclear transfer (SCNT) has met with almost complete rejection from scientists, philosophers, and the public, making it uniquely attractive fodder for the contemporary thriller. Popular novelists and filmmakers have been able to trust their sense of when readers and audiences will be captivated or horrified (or both) because there has been nearly universal agreement that this is one Pandora's box that should remain closed.

For that reason, the proliferation of genetic fantasy in the form of cloning narratives might seem easy to dismiss. What more fitting subcategory of popular fiction to endlessly reproduce than one that is itself about duplication? As we will see, though, the cloning tale has gradually become far more revealing than its superficial uniformity might suggest. Creating babies via cloning might remain a relatively unlikely prospect, but the stories featuring this possibility have become revealing challenges to common assumptions about the separability of nature and nurture. Just as scientists have demonstrated that human clones (like twins) would *not* have completely uniform genomes—a consequence of factors including the mitochondrial DNA in denucleated host cells—cloning tales are increasingly teasing out the interdependency of genetics and environment, and along the way, of self and other. While this capacity becomes most evident in the more realistic twenty-first-century narratives explored in chapter 3, its roots lie in print science fiction of the preceding century. Especially in their feminist incarnations, the genetic fantasies of the late twentieth century are an invaluable pre–Human Genome Project site for testing the nature of the soul and the potential for human beings to reach beyond traditional differences of class, gender, and sexuality.

Genetic fantasy's greatest strength might be in resisting the stultifying forces of homogeneity by openly acknowledging that abstractions like "unconditional love" and the "unique soul" are social constructions, but it might also be in insisting that they are *true fictions* that society cannot do without. Such nuance has sometimes been absent from Hollywood treatments of the clone figure as an inhuman blank slate, an empty form waiting for individual definition by the arrival of an immaterial soul. In contrast to the dualism of this model, some of the best print science fiction published between the late 1960s and the 1990s troubled oppositions not only of soul and body but also of self and other and of male and female. Beginning with Ursula K. Le Guin's 1968 short story "Nine Lives," this chapter examines one of genetic fantasy's most common masterplots, showing how the Carbon-Copy Clone Catastrophe both relies on and exposes corporatism's dependence on objectivism and conformism. Moving into subsequent decades, the chapter also demonstrates how cloning narratives offer increasingly nuanced challenges to sexism and homophobia. I look most carefully at Pamela Sargent's relatively neglected *Cloned Lives* and the better-known *Xenogenesis* trilogy of Octavia Butler (originally published as *Dawn*, 1987; *Adulthood Rites*, 1988; and *Imago*, 1989; and now collected as *Lilith's Brood*). Taken together, these cloning tales imagine an increasingly unbridled, sometimes explicitly queer form of love that recognizes the soul and its other as interdependent rather than opposed entities. In the process, they foreshadow genetic realism's even more explicit emphasis on the inseparability of genetics and environment. The chapter concludes with a jump forward to *Moon* (2009), a low-budget film that uncannily echoes "Nine Lives" and demonstrates how genetic fantasy continues to be capable of evoking both the strange idea of the human soul and its capacity for a weirdly unreserved love.

The Carbon-Copy Clone Catastrophe

We know this character. It floats in a bath of primordial liquid, unconsciously waiting to be awoken. White-coated geniuses outside the tank shuffle about, adjusting dials and furrowing brows, and the music swells in anticipation.

New life, new power: humanity has conquered the innermost terrain of the cell, hacked its own hardwiring. The body becomes clothing for a controlling consciousness, a soul independent of any specific embodiment. Think *Avatar* (2009), with Jake Sully transferring his mind from a paraplegic human body to a wonderfully unimpeded Na'vi one. That film might not be a straightforward cloning tale, but the floating, uninhabited corpus that awaits Jake on Pandora has appeared in science fiction for decades. Built to complement Jake's deceased twin brother's DNA, this alien avatar is the same kind of fully grown, soulless "blank" that awaited obedient suburbanites as far back in cinema history as *Invasion of the Body Snatchers* (1956), where it was encased in an enormous bubbly seed pod (fig. 6). *Blade Runner* (1982) never quite shows viewers one of these empty clone bodies directly, but they are visible over Rick Deckard's shoulder on a computer screen that displays generic images of various models. Like many filmic clones, Ridley Scott's "replicants" are born as adults, implanted with false memories of a previous life, and granted a very limited life span. Their descendants eventually complicate the marriage of Michael Keaton's character in *Multiplicity* (1996; fig. 7) and then Schwarzenegger's in *The 6th Day* (fig. 8); on the small screen, Cylon clones become virtually indistinguishable from "authentic" humans in the reboot of *Battlestar Galactica* (2004–9). For such beings, there is limited subjectivity or agency; their bodies are mere shells temporarily hosting an entity we only tentatively call a soul. Beneath all these genetic fantasies lie the most horrifying questions of all: Might the empty clone body *already* mirror its future occupant's soul, or even the viewer's? Is there actually any real presence that manipulates limbs and articulates speech, or is humanity's endless anxiety about personal identity merely the chemically induced product of wishful thinking?

These are abstract questions about the nature of the inner self, and most thrillers pause to consider them only briefly.[1] But Hollywood's cloning epics are not alone in suggesting that outward conformity to a society's expectations is insufficient for creating a strong sense of identity and belonging. Ironically, isolation is the dominant motif of cloning stories, not just figuratively, but usually quite literally, and not only on the big screen, but throughout the history of print science fiction. Some clones examined in this chapter, including those in Le Guin's "Nine Lives" and in *Moon*, are removed from other humans by the extreme distances of outer space. But even earthbound clones like those in the low-budget German films *Blueprint* (2003) and *Womb* (2010) or the mother and clone child at the center of John Case's novel *The Genesis Code*

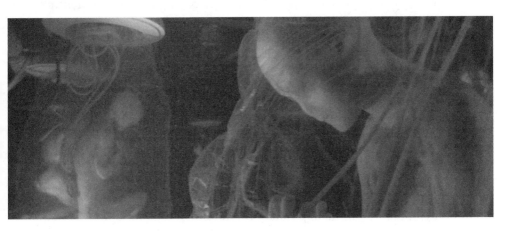

Figs. 6–8 Empty clone bodies in *Invasion of the Body Snatchers* (1956; fig. 6), *Multiplicity* (1996; fig. 7), and *The 6th Day* (2000; fig. 8).

(1997) withdraw to islands or remote coastal areas. When space or oceans do not provide seclusion, novels like Michael Marshall Smith's noir thriller *Spares* (1996) and Hollywood spectacles like *The Island* hide away their clones in other kinds of unvisited, inaccessible spaces, whether Smith's carefully guarded, windowless "Farms" or *The Island*'s hidden underground. These settings might encourage intimacy or hostility among the clones themselves, but in nearly every case, the clones are separated from the rest of humanity, living in a state of seemingly infinite longing, each fulfillment of which turns out to be fleeting, if not patently artificial.

The paradox is that things often appear completely opposite to the traditionally reproduced humans who observe the clones from afar. In one of the earliest examples of the subgenre, MacLean's "Diploids," even those who oversee the clones betray their envy, asking, "There must be a lot of pleasure in the idea of shaking hands with yourself and forming a mutual admiration society, huh?" Yet the short story's subsequent free indirect discourse sarcastically testifies to a different feeling for those on the inside: "Brothers closer than brothers, fellowship and understanding to end the loneliness of being different and separate and unable to join wholeheartedly with the people around you. Loneliness can become so basic that a whole personality is built on it. Who would know better than a diploid?" (56). Certainly not those who control and routinely exploit them. Even when it seems a nonclone character develops empathy for these entities' strangely other-less existences, as does Deckard in *Do Androids Dream of Electric Sheep?* and *Blade Runner*, that person's human status must be called into question. In Dick's novel, Deckard becomes so "conscious of his own aloneness" that he finds himself awkwardly defending rather than destroying Rachael, as his job requires (154). When she expresses fear that "we *are* machines, stamped out like bottle caps. It's an illusion that I—I personally—really exist; I'm just representative of a type" (165, emphasis in original), he somehow decides that "legally you're not [alive]. But really you are. Biologically. You're not made out of transistorized circuits like a false animal; you're an organic entity" (173). This blurring of the line between artificiality and authenticity does not stop with Rachael, as Deckard's own ability to prove his humanity comes into question. Scott's loose film adaptation goes so far as to have the bounty hunter help his artificial charge escape, having faced strong circumstantial evidence that he too is a replicant relying on implanted "memories."

Le Guin was already grappling in 1968 with this cycle of misunderstanding, envy, and self-rediscovery. "Nine Lives" lies near the front edge of a significant expansion of cloning tales beginning in the late 1960s, a period marked by increased concerns about the abuse of governmental authority, especially after Watergate in 1972. Geneticists in the United States were worried enough about safety to impose various moratoria at the 1975 Asilomar Conference on Recombinant DNA.[2] This gathering would be referenced often in the following decades as a key moment in the history of bioethics, as researchers, clinicians, and lawyers agreed on a set of guidelines for appropriate application and containment of new possibilities for genetically modifying organisms. However, this period is just as significant for the many cloning tales it sparked, narratives that continue to heavily shape public attitudes toward genetics. Fusing historical circumstances with the long-established literary and cinematic myth of the double or the doppelgänger, cloning stories were everywhere in late twentieth-century popular culture. Ostensibly fueled by the headlines of popular science, they were also driven by much deeper and more enduring questions about individual uniqueness and the extent to which having a body is taken as evidence of personhood.

Many genetic fantasies rely particularly on the Carbon-Copy Clone Catastrophe, a masterplot in which apparently identical clones are created and deployed en masse, only for disaster to ensue. Applicable to *Where Late the Sweet Birds Sang*, Ira Levin's *The Boys from Brazil* (1976) and its film adaptation (1978), the *X-Files* "Eve" episode, *Multiplicity*, *The 6th Day*, *The Island*, and many other novels and films of the last half century, my neologism plays up these stories' scientific absurdity but also suggests how their true significance lies outside technological speculation. It has long been a common assumption that cloning produces literally undifferentiated organisms, and tales like these have perpetuated that illusion. The laboratory reality, though, is that cloning a tabby cat can quite easily end up producing kittens with both striped and solid coats; there are more factors at work even in such a simple matter as coat pattern than are determined by the DNA of a cell's nucleus alone, including environmental factors during gestation. Nonetheless, these stories remain significant for the way they challenge some of childhood's most sacred lessons ("your fingerprint is absolutely unique," "there is no one in the world who is just like you") and for their critiques of rhetoric about the soul that seems designed only to bring false comfort to the masses. "You," in this disturbing formulation, become nothing more than the evolving expression of biological wetware, a design that is eminently replaceable.

To be clear, this is not what the biology of human cloning would necessarily suggest, nor what all these texts ultimately imply, but it is a fear many of them exploit. And this myth of exact duplicability is closely related to a second characteristic of many cloning tales. More than explorations of any literal potential for the loss of human differentiation, stories like "Nine Lives" are allegories about already present social injustice and class-based exploitation. In that sense, they sometimes craftily overturn their own premises. While allowing her clonal protagonists to remain exact duplicates in appearance, for instance—at least until experience begins to leave its scars—Le Guin shows that life in a community of those who are apparently "same" is far from the relational utopia it might appear. Le Guin meditates especially on the disorientation that ensues when a male clone whose only relationships have been with four male and five female mirror-beings suddenly must face existence as a singular entity. Notwithstanding the fact that *Playboy*'s original publication of the story illustrated all the clones as female, domino-like objects and neutered its author by initializing her first name, the tale remains a profound demonstration that the possibility—even the threat—of difference is a prerequisite for authentic relationship.

"Nine Lives" begins with a suggestion that its planetary setting is already imitating the deceptively uniform clone characters Le Guin soon introduces: "She was alive inside but dead outside, her face a black and dun net of wrinkles, tumors, cracks" (453). This is much the way Pugh and Martin view the seemingly indistinguishable characters who arrive to relieve them of duty at their lunar mining station. Readers approach the plot through the perspectives of these traditionally reproduced humans, whose daily lives entirely lack the glory associated with Apollo-era astronauts. At first, Pugh and Martin cannot begin to tell apart the five men and five women that make up the "tenclone." As Pugh observes two clones unconsciously reaching to brace each other as their transport stops, his jealousy is palpable: "Your skin is my skin, Pugh thought, but literally, no metaphor. What would it be like, then, to have someone as close to you as that? Always to be answered when you spoke; never to be in pain alone. Love your neighbor as you love yourself. . . . That hard old problem was solved. The neighbor was the self: the love was perfect" (459). Pugh's initial disquiet quickly hardens into disgust; when he wakes one morning to find that another pair of the clones has slept together, he can only view their relationship as incestuous. At the same time, though, he has to admire the tenclone's efficiency, its capacity to satisfy its own needs: "A clone, he thought, might indeed be the first truly stable, self-reliant human being. Once adult it would need nobody's help. It would be sufficient to itself physically, sexually, emotionally, intellectually. Whatever he did, any member of it would always receive the support and approval of his peers, his other selves" (461). At this point in the story, such a homogenous relationship might seem alien, but Le Guin has also suggested its attractions.

Even before the tragic accident that kills nine of the ten clones, however, Le Guin begins to test the assumption that the clones really are indistinguishable duplicates. As Pugh and Martin prepare the clones to take over their responsibilities, they gradually begin to note subtle disparities. The daughter of one of the United States' most famous anthropologists, Alfred L. Kroeber, Le Guin has always been deeply sensitive to cultural influences, so it would have been surprising if her story had left untroubled the carbon-copy myth by which clone bodies most often appear in fiction. Eventually, Pugh and Martin learn to tell two clones apart, "Zayin by gestalt, Kaph only by a discolored left further fingernail, got from an ill-aimed hammer at the age of six." Then the narration reflects, "No doubt there were many such differences, physical and psychological, among them; nature might be identical, nurture could not be" (461). This nature-nurture binary, of course, has long been a key tool for discussing developmental influences, but soon after this point, the story starts to suggest the categories' inseparability. Decades before Donna Haraway's vision of natureculture or Regenia Gagnier's theory of symbiology, Le Guin was imagining not only how nature and nurture both contribute to individual behavior but also the possibility of deconstructing the binary itself.

This intervention grows clearest after the break in the plot, as Pugh's and Martin's relationships with the clones change dramatically when the rumbling she-planet erupts against its exploitation. To that point, Pugh and Martin had found the tenclone an impenetrable society, polite but devoid of need: "Part of the difficulty was that they never really talked to Pugh and Martin. They joked with them, were polite, got along fine. They gave nothing" (461). Pugh had even begun to understand why, thinking to himself, "Why should they have sympathy? That's one of the things you give because you need it back" (462). Still, despite referencing their colleagues with a pejorative "it" (462), when the earthquakes interrupt radio contact, Pugh and Martin go after the clones, knowing they had been working near a mine entrance. Only one of the men has survived, and it takes all his colleagues' determination to help him survive the night as he relives the deaths of his siblings. Kaph's crisis is not primarily medical; rather, he is petrified by his sudden isolation and the necessity of truly facing others. As Pugh reminds Martin, who is somewhat less patient, "He never had to see anyone else before. He never was alone before. He had himself to see, talk with, live with, nine other selves all his life. He doesn't know how you go it alone. He must learn. Give him time" (469). The clone, too, has cultural and environmental conditions that must be considered, not just genetic inheritance.

Le Guin presents this vulnerability as the narrative's central test, both of its characters and her readers. Pugh persuades Martin to give Kaph a completely undeserved grace, and yet as we discover, this is possible only because Pugh identifies with the clone's plight more than he lets on. By the story's conclusion, it becomes evident that the clones aren't the only bioengineered humans in this story: Pugh himself has an artificial lung and corrected myopia. The former, he tells Martin, is "human, grown in a tank from a bit of somebody; cloned, if you like. That's how they make replacement organs, the same general idea as cloning, but bits and pieces instead of whole people. It's my own lung now, whatever" (470). Le Guin's move here is remarkable: even while the Carbon-Copy Clone Catastrophe was in its infancy, she broke down the clone-human binary and suggested how artificially reproduced beings might exist along a spectrum. The implication of Pugh's words is that they should treat Kaph as a full person because the scientific, material, and ethical gaps between artificial and original human body components are meaningless.

For Le Guin's story, in other words, the problem is far less with Frankenstein's creature than with those who fear him, and cloning begins to seem merely another routine form of biomedical intervention. Without the benefit of today's knowledge about the imbrication of genetic inheritance and environmental influence, Le Guin was already questioning readers about the rationale for separating the human body and its surrounding culture. Tellingly, the story ultimately centers on the capacity to love, something Pugh slowly teases out of a man who has never known the need. As Kaph hesitatingly asks, "How can you . . . How do you . . . [love someone?]" (475). Le Guin is far too artful to answer directly, as the whole story is her response. But her narrator concludes, "What can you do but hold your hand out in the dark?" (475). The final sentences are neither sentimental nor unfeeling: "Kaph looked at him and saw the thing he had never seen before, saw him: Owen Pugh, the other, the stranger who held his hand out in the dark. 'Good night,' Pugh mumbled, crawling into his sleeping bag and half asleep already, so that he did not hear Kaph reply after a pause, repeating across darkness, benediction" (476).

The simple reverence of this closing blessing is no more an accident than the clones being designated by letters of the Hebrew alphabet, even as the initials of their collective name, "John Chow," offer a reference to Jesus Christ. Nor is it a coincidence that Pugh restores Kaph to life using the "Lazarus Jab" (466). I do not mean Le Guin is attempting some sort of Jewish or Christian allegory, nor would I suggest that this is the story's effect despite her intentions. That would contradict everything we know about her lifelong affinity for Taoist philosophy, not to mention her disdain for propaganda, whether on behalf of a religious tradition or any other ideology. In fact, Le Guin's work regularly confronts Western culture's monotheistic fundamentalisms, from "The Ones Who Walk Away from Omelas" and *Always Coming Home* to *The Telling* and "Paradises Lost."[3] What I am suggesting, though, is a postsecular approach to "Nine Lives" and other genetic fantasy, one that reveals how heavily shaped its characters and implied audience are by religious language yet also resists attempts to cram the story's entire significance into a single ideological shape. Beyond any concern with the literal future possibility of human cloning, this tale is simultaneously a compressed bildungsroman, a critique of groupthink, and an invitation to see the virtues of difference, especially where it is least obvious. "Nine Lives" shows how the possibility of relationship depends on encountering a true other and then slowly building the harmony that Kaph discovers in a recorded chorale, with "a hundred human voices singing together" (470). As Le Guin's tale indicates, religious trappings can impede or enable the hard, slow work of welcoming the alien other, and sometimes, they do both at once.[4]

Technotranscendence and the Safety of Clone Sex

"Nine Lives" represents only the front edge of genetic fantasy's treatment of the clone figure, and it is far from unique in invoking religious tropes. In the 1970s, the clone's problem of isolation and loneliness was often presented as a crisis of religious faith, both individually and culturally. Cloning novels from the period routinely take on theological dimensions by presenting the work of science as an expression of religious dedication. Nancy Friedman's *Joshua Son of None* (1973) treats cloning as a literal means of resurrection, predating Stephen King's alternate history *11/22/63* (2011) in imagining how John F. Kennedy might have been brought back to lead the United States after his assassination. The novel's hymn to technotranscendence begins early and never lets up: "Genetically he would rise. The nightmare of death was over. Man could live again and again, forever if he wished" (9), and the inspired researcher who would make it possible "had belief too, and faith. His litany could be read under the electron microscope.... His mass, like the great B minor, was a celebration of life, but in new terms" (11). The wife of this JFK replacement tells her husband that the voters "see you as him [his predecessor], through the miracle of science restored to them. Some even see it religiously, as the hand of God working to restore what should have been" (195). As a result, even when the clone is struck down at his own inauguration, the scientist who created him only sighs and prepares to repeat the process, with the text repeating the free indirect discourse near its beginning: "Man could live again and again, forever if he wished" (211). As Friedman's popular novel joins many others in illustrating, such defiance of mortality has sometimes threatened to profane divine authority, but it has also been figured as fulfilling a divine plan. Doggedly, scientism returns to its test tubes, determined to let no ending prove final, a devotion it scripts as a naturalistic replacement for a previously religious mythology.

In the previous section, we saw how "Nine Lives" comes down to a male clone who must face life anew without his siblings; likewise, Friedman's story revolves around the potential rehabilitation of a male messiah. The clone's default masculinity is apparent again in Ira Levin's *The Boys from Brazil*, in which case it is Hitler whose personality might be reincarnated. Since the 1970s, in short, readers have been invited to identify far more often with male clones than their female siblings, especially in cinema. This pattern made it possible for less informed science fiction readers and movie audiences to marvel at the seeming innovations of Suzanne Collins's *Hunger Games* trilogy (2008–10), Veronica Roth's *Divergent* series (2011–13), and their film adaptations, in all of which relatively strong female lead characters battle against the biopolitical power of oppressive oligarchies.[5] However, such heroines have been central to print SF since at least the New Wave, and in many cases, their presence has retooled cloning narratives and particularly their usage of religious tropes. By calling attention to the body and its representation, feminist science fiction has made sexuality and gender issues integral to the cloning tales that reach furthest beyond easy religious-secular binaries and one-sided treatments of biotechnology.

Of course, just because a narrative features feminist insights does not mean it also cultivates postsecular thinking or wisdom about science. While cloning tales by and about women have often aggressively questioned genetic determinism, they have not always succeeded in offering an alternative. Like other genetic fantasies, the Carbon-Copy Clone Catastrophe has both resisted and succumbed to Watson's famous suggestion that fate is largely written in our genes. And even where these narratives have undercut easy formulas of genetic determinism, they have frequently remained vague about relationships among biological inheritance, environmental influence, and individual agency. Fay Weldon's *The Cloning of Joanna May* (1989), for example, represents both satire's power and its sometimes limited efforts to envision a better world. In a fantastic mode echoing Levin's *The Stepford Wives* (1972), which uses robotics instead of genetics to make its allegorical points, Weldon's novel sharply criticizes paternalistic biopower. Joanna objects that her geneticist husband "could snip out the section that decrees I will have long and elegant legs, and snip in a section from someone else's DNA, someone with short piano legs, the kind without ankles. He could give me a dog's back legs" (157). Such absurd modifications herald the downfall of religious power and cultural stability in Weldon's storyworld, as "the Holy Ghost flew off, muttering, 'Christ, where is this going to end?' for it's been trouble ever since, hasn't it, all downhill; war, and riots, and crime, and drugs, and decadence, and dereliction, and delinquency, because there's no God" (155–56). Still, the novel suggests no real substitute for the facile notions that genes are fate and that to modify them is divine. On the final page, Joanna remains stuck in an endless circularity much like that imagined by *Joshua Son of None*: she can express bitterness about her immobility but not escape it. Speaking for herself and her clone daughters, she concludes, "We would have been perfect people if we could, but our genes were against us. We would have been faithful, kind and true, but fate was against us. We are one woman split five ways, a hundred ways, a million million ways" (265). There is righteous anger here, but still only carbon copies.

Wilhelm's *Where Late the Sweet Birds Sang* goes further yet still hesitates to conjure a real alternative to genetic determinism. Like many cloning tales of the late twentieth century, Wilhelm's novel suggests a clone *should* be treated as a unique person but cannot quite imagine how. As with "Nine Lives," the story moves from casting doubt on the soulfulness of its clones—a main character notes that they seem to be "rejuvenated with something missing" (49)—to a nebulous hope for a future in which they can live as distinct individuals. The story begins with an ecological and social collapse echoing Noah's flood and a family of scientists using cloning to perpetuate the species, a masterplot more recently adapted by Margaret Atwood's *MaddAddam* trilogy. Years after the cataclysm, a character named David watches three mirror images of his deceased lover walk by, and his internal debate encapsulates the novel's dilemma about human clones: "Three

Celias came into view, swinging easily with the weight of the baskets, a stair-step succession of Celias. He shouldn't do that, he reminded himself harshly. They weren't Celias, none of them had that name. They were Mary and Ann and something else. He couldn't remember for a moment the third one's name, and he knew it didn't matter. They were each and every one Celia" (43).

As the passage's hairpin turn demonstrates, the novel's clones remain "indistinguishable" despite the best intentions of Wilhelm's observer (55). The novel does include an artistically gifted character who can see past the clones' outward shapes to "how different they were one from another" and even observes that "there is a slight difference between the babies cloned in the laboratory and those born of human mothers (71). There is the prenatal influence and also the birth trauma that might alter the sexually reproduced person" (131–32). However, *Where Late the Sweet Birds Sang* is ultimately unable to upset the carbon copy delusion because it ignores the impact of diverse experiences after birth or the power of personal agency. Most tellingly, we never actually meet two clone characters who can be told apart. Notwithstanding the novel's hasty epilogue, which informs us of a new settlement in which "no two of [the clones] were alike" (206) and in which the new focal character is happy "because all the children were different" (207), Wilhelm does not zoom close enough to describe any actual distinctions.[6]

For all their pleasures, the limitation of many 1970s and 1980s cloning novels, including those resisting traditional cultural norms, is that arriving at a situation in which apparent human copies emerge as unique beings seems to necessitate immediate narrative closure. This remains true in much blockbuster cinema, from *The 6th Day* to *The Island*: the story's big reveal is the carbon copies' discovery of their origins and a unique sense of identity, but that is where things must end. Only rarely do cloning tales consider from their beginnings that individuals with nearly matching genomes might still possess ordinary thoughts, feelings, and relationships. A significant early exception, however, is Pat Sargent's *Cloned Lives*. Rather than following Wilhelm's tack of aligning the reader with "normal" protagonists who watch the clones with discomfiture or envy, each of Sargent's seven sections focuses on one of the novel's five clones, their original, or the entire clone family. Indeed, a "normal" family is what the novel considers these characters, with all their distinct talents. While their genetic "father" (another Paul) is an astrophysicist sometimes tempted to think of his infant children as exact replicas of his earlier self, he is wise enough to find teachers and nurturers for the children who will "treat them as individuals, not just as a group" (53). As a result, while Maria Ferreira understandably reads the novel as illustrating Baudrillard's "hell of the same" (147), *Cloned Lives* stands out for its commitment to revealing individual character behind superficial similarities. Sargent's clones' personalities and interests diverge as much as those of other human siblings: Edward is most comfortable when alone, preferably with a calculus problem; James finds it torturous to be a clone, and after nearly committing suicide becomes a rebellious novelist; Michael is a people-pleaser who loses himself in electrical engineering and physics; Kira, a biologist since birth and the quintessential peacemaker, studies the ethical issues surrounding their conception; and Albert's path leads to the moon, where he works at an astronomical observatory.[7] Though more genetically similar than twins, neither the clones nor their progenitor become the human monsters on which cloning tales often rely. They are aware of others' speculation about "mass minds, or mental telepathy among clones" (66), and James worries about perceptions of them as "identical, a closed group, undifferentiated, and inaccessible" (122). However, Sargent explodes the mythology here, treating anxiety about being overlooked in a sea of homogeneity as a more universal human problem.

Not by chance, Sargent's only female main character serves as the novel's major voice of insight. Rejecting the habit of self-ostracism familiar to many clone tales, Kira demands, "Are any of us so unusual? Don't people all have the same roots anyway? No one's an isolated self, we're all different really, but that doesn't mean we have to isolate ourselves" (128). By aligning the reader's perspective with the clones rather than with traditionally born characters, Sargent invites readers to recognize how many human beings fear being reduced to a function. Michael's reaction when the siblings are finally reunited on the moon is emblematic of more than just clone lives: "He suddenly hated them all, hated them for being his relations. He found himself slipping into the role of sibling again, having to spend time with people who had nothing in common with him except genes" (297–98). If this genetic fantasy is more perceptive than most in rejecting the Carbon-Copy Clone Catastrophe, that is because it is also about life within any family.

Sargent also uses her clones to resist the soul's assimilation within assembly-line capitalism. A character in a bildungsroman rebelling against fatalistic pressures is nothing unusual, but *Cloned Lives* both literalizes and resists this sense of inevitability. Its first sentence highlights the problem with repetition that often consumes cloning tales: "As the jet approached the Dallas–Fort Worth Regional Airport, Paul Swenson saw the nearest of the circular loops which made up the huge, monotonously efficient structure" (12). As in Le Guin's story, society's manufacture of homogenous structures concretizes authority and impedes personal growth. Sargent portrays James as the most sensitive sibling to this threat, and it leads him into depression and nearly suicide. As if recapitulating his father's life, "He felt paralyzed. He saw himself as a puppet, walking through an ever-repeating cycle. *I'll go through it again*, his mind murmured, *I'll go on feeling the way I do, acting the way I do, and I won't have any choice. It's all happened before and I have no way of changing it*" (139, emphasis in original). Even relatively late in the novel, he tells Kira as they bid each other good-bye, "I think we've played this scene before" (232), and she agrees. The existential complaint of being stuck in an endless cycle extends far beyond the fantasy of carbon-copy clones, but Sargent uniquely exploits the tautology inscribed in their very gestation.

Cloned Lives's significance, then, is that even as its characters struggle with a sense of determinism, readers can see that it is misleading, that these characters' strategies for responding to the cultural expectations under which they are placed are as diverse as their interests. Even if the novel sometimes draws unnecessarily tight boundaries around the potential expressions of that diversity, Sargent's achievement is most visible in these characters' responses to sexual desire. The novel rejects the clone's common treatment as the ultimate blank slate and wrestles openly with what it means for love to require otherness. In Kira's relationships with her clone brothers James and Edward, for example, we see how she both stands in for the clones' nonexistent mother (the ultimate source of familiarity) and uses her physical allure to provide affirmation (as if she is uniquely capable of welcoming their differences). Most strikingly, their intraclonal sex takes the incestuous element that revolts Pugh and Martin in Le Guin's story and presents it from the clones' points of view as a first satisfying and then insufficient expression of longing.

On the surface, what we observe is sexual intercourse between opposite-gender clones, which could seem a reminder that the clones vary in at least one way. However, that is not the language with which the novel figures these encounters. Despite its empathy for the siblings' isolation, these encounters come across as masturbatory and insufficient. James overcomes his hesitation about Kira's advances, we are told, not with an eye to her differences but by deciding that "she was his female self." His desire is not an expression of openness to an other, despite its gender-crossing valences, but a narcissistic fantasy of self-satisfaction: "He saw himself as a woman, receiving a man, opening to the hardness that plunged inside her, and knew that she was seeing herself as a man." Initially, the novel seems to idealize the experience, certainly when compared to Le Guin's description of Pugh viewing clone sex as incestuous: "*He saw millions of men and women seeking mates, trying to find those who would complete them, make them whole again, yet always separated from them by the differences passed on to them by eons of change. He saw Kira and himself, reflections of each other, able to move along their individual paths and yet meet in perfect communication. She was no longer his sister, but his other self, closer to him than a sister could have been, merging with him so completely and perfectly that they were one being*" (145, emphasis in original). Soon after celebrating their lovemaking, though, James reinterprets it as a surrender to genetic determinism and his fear of risks. In effect, Sargent seems to affirm the reactionary judgments that Le Guin's story exposes in her character Pugh. While "it would be easier to stay with Kira, easier to give up on other people," James decides "he would not let himself do it yet, not until he had tried and failed many times" (150). As James never withdraws this commitment, the novel ends up equating intraclonal sex with narcissism.

This conclusion unnecessarily falls back toward treating clones as carbon copies and condemns queer desires. The details of the scene and its psychosexual denouement make the novel's ambivalent tone particularly striking. As James comes together with Kira, he is focused on his absent father, evincing a mutated Oedipal complex that provides reconciliation with the father rather than usurping him: "It was Paul's face that watched him, smiling, gently reassuring him, protecting him with love." If that isn't provocative enough, the next section's treatment of Kira flirting with another of her clone brothers, Edward, insists on a Freudian reading: "Kira hooted as she aimed the hose at Ed, drenching him completely. He grabbed the hose from her and began to spray her with water. Kira danced on her toes, laughing loudly" (146). Then Kira and Ed head to the house for their own round of incest, leaving James outside, where he "finished spraying the last chair and glared at the hose. He was annoyed that Ed and Kira had not rewound it and put it away; it was not like them" (147). Galled by this untidy, dangling signifier, James follows the two inside and interrupts their lovemaking. His response is to depart, realizing he cannot give up on relationships outside his family but must keep trying to build bridges across difference—which presumably cannot exist among clones.

There is a similar reversal in the novel's ultimate embrace of technotranscendence and hyperefficiency. While its first sentence called attention to the capitalistic utopianism that undergirds much rhetoric about genetic engineering, the novel's resistance finally erodes. In the first chapter, readers meet an unconventional clergyman who worships "the universe itself and the principles behind it" and speaks of "Jesus as an example rather than as the Son of God," and one senses the story might begin to question traditional religious-secular binaries (29). However, this figure quickly disappears from the novel, which in its second half falls back on Cold War conflations of technological achievement and religious fulfillment. A favorite science fictional novum of the period, nuclear fusion, becomes not just a means of speculating about future energy independence but a trope for the novel's entire treatment of the relationship between matter and spirit. If fusion is the corollary in nuclear physics to cloning in genetic biology, *Cloned Lives* succumbs to the utopian dreams of its characters: "If [Mike] and his colleagues could perfect their matter-scanning techniques, have an image of every atom's place within an object, they could, with the energy released by fusion, duplicate almost anything. Every person on earth might eventually have access to any material goods he desired or needed. . . . The day was not far off when all the powers of the world would be equalized, when their greed and material comfort, cemented by a humane technology, might let them at last trust each other and work together on the next set of problems that would confront humanity" (173). Endless repetition, Sargent's novel posits, might finally solve the problem of greed not by confronting it but by satisfying it.

Rather than distinguishing between the methods of science and the metaphysics of scientism, then, Sargent's novel finally joins many other thrillers in letting one slide into the other, just as its initial critique of homogeneity drifts into an implicit homophobia. The story's conclusion confirms this descent, with the clones bringing back Paul's cryogenically preserved body from death and any remaining connection to the novel's early bioethics completely disappearing. The future imagined by *Cloned Lives* might be imperfect, but the novel ends up waxing poetic about escaping death. Kira only briefly hesitates, since death had provided an "ever-present certainty that everyone, high or low, famous or forgotten, would have the same end, that they would all become equal in the grave" (306). Now though, "she saw another humanity on Earth, freed from the determinants of genetic disabilities, of aging bodies, of unbalanced minds, and of death" (308–9). With this blithe vision of future utopia, the novel reaches beyond human cloning toward an era in which birth and death fully surrender to human manipulation. In his resurrection, "Paul had become something new in the world" (326–27), and we last see Albert looking boldly into a technotranscendent future aboard a multigenerational starship: "The ship was a prayer to the universe, a request for its secrets, the embodiment of an enterprise based largely on faith. But the priests and priestesses of this shrine would not wait for revelations; they would actively seek them. Their chants and holy words would utilize the power of mathematics, observation, and physical laws" (330). Launched by a society that has conquered death—and in the process rendered sexual reproduction unnecessary—Albert's journey will mirror Michael's work in fusion. Both clones will throw off the shackles of their earthly, bodily limitations, and neither will grasp how much their culture's sexual hierarchies have tightened those restraints. In short, Sargent's novel represents the cutting edge of 1970s cloning tales in its attention to environment and individual choice, but it finally capitulates to the strictures of heteronormativity and a vision of biotechnology as a passageway to the heavens.

Queer Clone Love and the Dawn of Fictional Epigenetics

A decade later, Octavia Butler's *Xenogenesis* trilogy indicates the potential of genetic fantasy to go further. Set against Sargent's tale, Butler's three novels feature a steadier sense of the power and danger of technotranscendence, a far more hauntingly attractive sense of the alien other, and a much less fearful approach to queer sexualities.[8] At first glance, *Dawn*, *Adulthood Rites*, and *Imago* might not constitute an obvious cloning narrative, and Butler only rarely uses that term. However, once a reader steps back from the trilogy and reflects on the extent to which the alien species of the Oankali and particularly their third sex, the "ooloi," are committed to obtaining genetic "prints" of both human beings and other organisms, the connection becomes clear.[9] The premise of this acclaimed African American writer's most famous series is that via World War III, humanity has already destroyed itself once with its dangerous combination of intelligence and hierarchical behavior, and unless its biology can be rewritten through relationships with an alien species that is simultaneously its colonizer and its rescuer, it will likely do so again. The principal hope these novels allow comes via genetic engineering, yet Butler means much more by printing or rewriting an individual's genetic code than do today's laboratory geneticists. For *Xenogenesis*'s somewhat Lamarckian concept of evolution, genes transmit not just biological inheritance but also cultural memory—indeed, these data are inseparable. Fitting its Gaia-like vision of a spaceship as a living, responsive organism, the trilogy posits that to be a soul is to live in interdependent relation.[10] Beyond confronting fears of otherness, Butler's work cultivates an active hunger for the unknown. The novels imagine radical embraces, both literally across alien species and figuratively across differences of race, gender roles, and sexual preference. Butler's work is as queer as the twentieth-century cloning tale gets, not to torment traditionalists or to titillate adventurers, but because a deep heterogeneity is inextricable from her complex, sometimes very troubled vision of selfhood and relationship.[11]

The first novel, *Dawn*, focuses on Lilith Iyapo, a dark-skinned woman who awakens to find herself aboard an alien ship. She learns that her rescuers from the remains of a nearly global nuclear holocaust are the Oankali, aliens who "trade" genetic information to attain greater diversification. Like the other humans that Lilith is eventually called on to awake, she grasps her situation only gradually. At least temporarily, she must trust Jdahya, the first Oankali she meets, and this becomes even more necessary as she recognizes the resulting distrust with which her fellow humans regard her. Since she has been given genetic enhancements to manipulate the ship and to defend herself physically, Lilith's comrades are dependent on her but also suspicious. She is set up from the very beginning—ironically, given the Oankali's professed aversion to hierarchical relationships—to function simultaneously as prophet and scapegoat (Butler prefers the term "Judas goat"). As a result, after the murder of Joseph, the man Lilith took for her partner, Lilith is surprised to find herself closer to the Oankali who seduced her than to other humans. Like a female Moses, instead of being allowed to lead her charges into the earthly Promised Land, Lilith is required to stay behind with her "ooloi" Nikanj to awaken another group on the spaceship.[12]

Dawn is one of the most widely teachable of science fiction novels, not just because of the attractive simplicity of Butler's plot and syntax, but because its opening scenes so grippingly portray an encounter with apparently absolute otherness and because its later chapters so effectively dramatize a range of human responses to disempowerment. Butler begins by forcing her main character to endure nearly complete solitude over what she experiences as a roughly two-year period; manipulated like a zoo animal, Lilith is denied the most basic information about her circumstances. Some interpreters hear resonances of antebellum slavery in these scenes, but Butler is writing even more broadly about colonialism and cultural assimilation. While she begins with a human perspective in *Dawn*, the later novels' shifts in focal characters and the growing ambiguity with which she treats the possibility of genetic determinism make the treatment far from one-sided, as we are challenged to empathize with human and alien alike. Indeed, one of the trilogy's central questions is how cultural habits of domination might function as *part* of biology, neither above nor below genetic inheritance. When Jdahya explains that the Oankali's dual attraction to and revulsion from humanity is due to the species' unusual combination of intelligence and hierarchy, he adds, "It isn't a gene or two. It's many—the result of a tangled combination of factors that only begins with genes" (39). Butler not only alludes here to pleiotropy (a trait's influence by multiple genes) but also indicates a deep awareness of how gene expression depends on environmental context.

That does not make these novels genetic realism; they are not heavily concerned with how best to understand and respond to contemporary biotechnology.[13] Butler's real interest is in the functionally colonialist, pseudodivine position of a relatively omniscient, hyperadvanced society of bioengineers relative to their (re-)creations. How might one build healthy relationships across such a massive gap in power, whether literally between species or allegorically between races, genders, classes, or nationalities? One early signal of Butler's fantastic orientation is Lilith's sudden knowledge of the Oankali language on awakening from one of Nikanj's genetic manipulation sessions. Words appear here as simple units of precise cultural code in the same way that genes are often popularly misunderstood as simple units of precise biological knowledge: "It was even easy for her to spot the gaps in her knowledge—words and expressions she knew in English, but could not translate into Oankali; bits of Oankali grammar that she had not really understood; certain Oankali words that had no English translation, but whose meaning she had grasped" (81). The further one reads in the trilogy, though, the more these seeming equations of signifiers and signified and of genes and traits become confused. The trilogy emphasizes that all the biological matches in the world do not necessarily create complete identification of the self with its clone, nor between the genetic colonizer and the colonized. Sometimes Butler lets go of this perspective, as when Nikanj explains to Lilith early in their relationship that if she were cloned centuries into the future, "your body might be reborn" but that "it won't be you. It will develop an identity of its own." Still, only a few sentences later, Butler comes right back around and allows Nikanj to speak of this "gene map" as "a plan for the assembly of one specific human being: You" (99). In other words,

Dawn senses the tension between genetics and personal destiny but struggles to describe it. Butler believed with Nikanj that humans are "more than only the composition and the workings of your bodies," that their identities are just as much a function of "personalities" and "cultures," but like many geneticists in the late 1980s, she was still unsure of how to represent that excess (153–54).

Adulthood Rites, the trilogy's second novel, features additional speculation about a genetic reality beyond contemporary science's discoveries. This time, the novel's main character, Akin, is a male Oankali-human hybrid, a product of genetic material contributed by male and female human parents, male and female Oankali parents, and a third-sex ooloi, an Oankali parent who engineers the mixture within its body. Akin's abilities far surpass those of entirely human children his age, but they are not without limits. At root, he is an explorer who thrives on examining other persons at the genetic level. We are told that while investigating Lilith's DNA, he is frustrated to find "something beyond the nucleotides that he could not perceive—a world of smaller particles that he could not cross into" (257). Akin has a similar reaction later to a brief study of Lilith's eventual partner Tino, again suggesting how Butler was attempting to reach beyond genetic determinism without quite knowing what she would find. "He would have preferred to investigate further, to understand more of how the genetic information he read had been expressed and to see what nongenetic factors he could discover," we are told (273). It never quite becomes clear what these "nongenetic factors" are, but whether they influence race, sexuality, or some other aspect of identity, Butler's choices—like those of the Oankali—always favor hybridity and unpredictability, even if they also leave room for exploitation and assimilation. This might be why the Oankali allow some humans to escape immediate surveillance and begin life on a new planet, acting on the belief that diversification and heterogeneity are inherent goods. Given what we learn of the aliens' eventual impact on other species, it remains debatable whether this is actually a mercy.[14]

In any case, it should not be surprising that such choices invite theological discussions about determinism and freedom. Butler grew up the daughter of a Baptist minister, and her entire oeuvre is inflected by youthful experiences with Protestantism.[15] Her later *Parable of the Sower* (1993) and *Parable of the Talents* (1998) sequels might seem more directly relevant in that they explicitly imagine the evolution of a new religious movement, but their dictum that "God is Change" applies just as fully to the evolutionary, protoepigenetic vision of this trilogy. In fact, a key to understanding *Xenogenesis* is to see how it elevates uncertainty beyond a name for inert hesitation toward describing a more dynamic way of being. This might seem counterintuitive, given that what attracts the Oankali to humanity, more than anything, is *cancer*: what one species experiences as entirely destructive, the other sees as potentially constructive. Both literal and physical mutations in the trilogy's human, Oankali, and hybrid characters are unpredictable and sometimes very painful developments, but while they might sometimes cultivate fear and heighten suspicions, Butler's trilogy affirms them as the very spice of life. The Oankali are all but certain that a new start for humanity on an ecologically modified Mars would only delay the species' self-destruction, but Akin decides that no matter how likely that outcome, the opportunity is necessary. In demanding an option for humans beyond "either union with the Oankali or sterile lives free of the Oankali" (404), he points out the injustice faced by any people "freed" by a colonizer or enslaver but then denied the tools necessary to thrive in their new context. Instead, Akin puts significant hope in uncertainty: "Chance exists. Mutation. Unexpected effects of the new environment. Things no one has thought of. The Oankali can make mistakes," he muses (501–2). In short, this half-human, half-Oankali protagonist acts on the hope that no matter how much biology or anthropology one knows, there are dimensions of the future that remain unforeseeable. Even at the cellular level, events must play out in order to be truly known.

In considering Butler's mixed response to the lures of biotechnology, it is especially fitting that Akin inherits Lilith's bridge-building role between the Oankali and humanity. Maintaining simultaneously individual and group identities, he recognizes his family even as a fetus: "He was himself, and he was those others" (255). Later, he continues to identify with the humans he encounters, understanding their vulnerability and fears more fully than any other Oankali. It is for this reason that Akin has occasionally been recognized as a Christ figure, but unlike *Cloned Lives* or more recent thrillers like *In His Image*, Butler's trilogy does not worship at the technotranscendent altar.[16] It is difficult to discern "good guys" or "bad guys" in this genetic fantasy; rather than assigning blame, *Xenogenesis* recognizes how inherited mythologies can both hinder and assist our vision. Sometimes addressing their human charges from above, the Oankali play a divine role in the new gardens they establish, but their sensory organs are writhing appendages, a "nest of snakes" reminiscent of Eden's serpent (13). Like Yahweh, the Oankali clothe the humans they rescue, but until Akin's intervention, they also deny their charges any real independence. Most shockingly, the human couples who accept relationships with an Oankali ooloi as mediator find later that they are repulsed by each other's direct touch. In brief, Butler figures the "love" displayed initially by the Oankali as an ironic exercise in assimilation, objectification, and addiction that creates considerable friction with their rhetoric of salvation.[17]

Alongside this dissonance, however, the trilogy also imagines a new, unapologetically weird form of love that reaches beyond the most ingrained of biological and cultural barriers. This becomes most visible in *Imago*, where Butler's least-guarded treatments of queer desire appear. All along, we have known that the Oankali attraction to humanity is a compulsion they can barely control. Nikanj explains to Lilith in *Dawn* that "you've captured us, and we can't escape" (153), and in turn, she explains to Akin in *Adulthood Rites* how "Oankali crave difference" (329). What becomes clear in *Imago*, though, is that even if this addiction produces what appears to be "a dangerous mistake" (545) like Jodahs, the drip on the painting might be the moment it grows most interesting. One might even argue that Jodahs is the trilogy's ultimate Christ figure in that he experiences an even greater rejection than Akin due to his unanticipated birth. He wanders into the wilderness because he refuses to deny the extent to which he holds together not just the human and Oankali species but also male and female gender roles. To be sure, neither he nor any other character in the trilogy is a simple allegory for queer identities, but we should recognize the reactions he sometimes evokes. The first human healed by this homeless messiah can only respond with hatred and an insult that is suggestive of homophobia's often unacknowledged basis in sexism. Instead of dying alone, João finds himself growing a healthy new leg, yet he can only accuse his physician: "You treat all mankind as your woman!" (599). This is machismo exposed: an inability to accept vulnerability, a simultaneous need to demonize the female as temptress and to render her the passive victim of male violence.[18] Appearing feminine is the worst thing João can imagine, and his ejaculation is so violent that it is easy to forget that Jodahs has just made the lame walk, as he will later allow the deaf to hear.

The title of this final novel in the trilogy is particularly telling. What is Jodahs the "image" of? Not just of the divine, as in Genesis, but of humanity's interpenetration with the divine, as in Revelation. Jodahs is "the most extreme version of a construct—not just a mix of Human and Oankali characteristics, but able to use [his] body in ways that neither Human nor Oankali [can]" (549). He is something new, something unexpected, far more than the resurrected Paul in Sargent's novel. The word Butler chooses to describe Jodahs here, *synergy*, has been used theologically to describe the mutual participation of the human and the divine in regeneration, physiologically to describe the work of interdependent bodily organs, and more broadly to indicate any phenomenon by which a whole becomes greater than the sum of its parts.[19] All these connotations contribute to Butler's positioning of Jodahs as the future of not only humanity but being itself. We learned at the beginning of the trilogy that the term describing the third sex of the Oankali, *ooloi*, means "treasured stranger" (106), but in *Imago*, we further discover that it signifies "weaver" and "bridge" (526). In *Dawn*, Lilith was taken aback by Nikanj's "unbridgeable alienness" (97), but by the end of *Imago*, the earlier source of fear produces the means of reunion. This does not mean that Butler is merely coming full circle. She is not interested in mere repetition, in endless cloning without innovation or mutation, no more so than Jodahs is excited by life back aboard the ship, where the trilogy began. That is a "finished place" that lacks "wildness" and "newness," and if there is anything that the veiled, racially and sexually transgressive Christ figure at the climax of Butler's trilogy represents, it is that there can be no life without ongoing difference (613).[20]

The trilogy's final scenes are thus its closing statement on the value of heterogeneous combination, and names again play a significant role in suggesting Butler's purposes. The fertile humans whom Jodahs attracts are "Jesusa" and "Tomás"; the latter name, incidentally, recalls both the disciple who demands physical evidence of the resurrection and a gospel deemed too "other" for inclusion in the New Testament canon. As elsewhere in the trilogy, Butler tweaks the allegory by inverting their roles, with Tomás trusting Jodahs long before Jesusa does, but the revision of Genesis's encounter between Adam and Eve and Yahweh could not be more explicit. Jodahs woos the humans with "tubers," a sort of "applesauce fruit," and this time, the man eats first and then convinces the woman. The humans' choice to flee is also instructive: they are taken aback by the radically unrestrained sexuality and love that this "construct" god offers. Ultimately, though, they recognize that their mythology has been misleading—that the seductions of the Oankali might be overwhelming but the aliens have also been unfairly caricatured as "devils" (661). Intriguingly, Butler chooses to paint this last camp of fertile human resisters (then converts) as explicitly Christian, with "elders" crossing themselves, with characters named "Paz" and "Santos," and most importantly, with the assumption that to be "Christian" is to be good. Jesusa objects initially that the love Jodahs wants with them is an "alien thing," and in a telling equation, "an un-Christian thing, an un-Human thing" (648). There is a dualism of body and soul here that makes sexuality inherently shameful, but by the trilogy's conclusion, it is a mistake that both the Oankali and the trilogy as a whole powerfully contest.

In an interview with Larry McCaffery and Jim McMenamin conducted while Butler was finishing the novels, she made her insistence on balancing genes and culture quite clear. "To whatever degree human behavior is genetically determined," she said, "it often isn't determined *specifically*; in other words, no one is programmed to do such-and-such" (62). She reinforced that position in a 1996 conversation with Stephen W. Potts, acknowledging that "some readers see me as totally sociobiological, but that is not true. I do think we need to accept that our behavior *is* controlled to some extent by biological forces.... But I don't accept what I would call classical sociobiology. Sometimes we can work around our programming if we understand it" (332–33, emphasis in original).[21] This is the insight that generates the unique souls and strange attractions in *Xenogenesis*. Well before most writers of genetic fiction, Butler was convinced that the relationship between genetic inheritance and cultural context was far more complicated and shifting than it is often represented. Molly Wallace puts this particularly aptly in an article that concludes by rejecting oversimplifications of genetics in Butler's trilogy: "*Xenogenesis* thus works against any notion that the genome is actually a book, that genes are literally selfish, or that genes are naturally tradable" (124). Instead, Butler saw how dynamic gene-environment interactions lie behind every expression of love and hatred. Correspondingly, it is only possible to understand human behavior by treating race, gender, and sexual preference as simultaneously biologically underwritten and culturally constructed.

Genetic Fantasy on the Edge of Realism

In the next chapter, we will turn to genetic realism, a phenomenon that emerges during and immediately after the Human Genome Project. This was the result of a significant mutation in genetic fiction, whereby it grew more interested in scientific plausibility, character interiority, and structural complexity. This dating, though, should not be read as absolute, or as if genetic fantasy ended when genetic realism began. Genetic fantasy continues into the present, both in print and in screen thrillers. The *X-Men* are no less popular than they were in the 1960s or 1970s; one might argue that they are even more relevant now, especially given ongoing violence around race, class, and sexual preference. Conversely, by identifying a millennial surge in genetic realism, I do not mean to suggest that there were no earlier literary efforts to adjoin the conventions of literary realism and knowledge of genetics. From Huxley's *Brave New World* to Powers's *Gold Bug Variations*, there are exceptions. Nonetheless, in the 1990s and especially in this century, there has been a notable shift in genetic fiction's orientation to the present and near-term future, a pattern that has both reflected and cultivated a rise in public awareness of biotechnological possibilities.

To probe the sometimes hazy boundary between genetic fantasy and genetic realism, this chapter concludes with a 2009 film that functions cinematically as realism, but in its loose relationship to biological knowledge, it relies on genetic fantasy's allegorical treatments of the soul and its relationships for its narrative. Duncan Jones's *Moon* uses a combination of gritty, timeworn stage sets and intricate, low-tech models to create a vivid picture of the potential love between even the most programmed of human clones.[22] Sam is an astronaut miner living alone on the moon, entirely unaware of his genetic status or his limited life span. He supervises four enormous rock harvesters amusingly named Matthew, Mark, Luke, and John, as if the machines were the four witnesses to his messianic task. Finishing a three-year tour of duty, he expects to return home to his family soon, but a rover accident leaves him unconscious and stranded on the moon's surface. As a result, Gerty—a warmer, more sympathetic update of the HAL 9000 artificial intelligence in *2001: A Space Odyssey* (1968)—decides to awaken the next Sam clone (hereafter Sam_2). Alongside Sam_1, who is rescued from his crashed rover by Sam_2, the film's viewers gradually understand that moon base Sarang, or 사랑 ("love" in Korean), is the corporate altar on which an endless supply of clones is being ritually sacrificed. Every three years, a new Sam is awakened to learn from Gerty that he's had an accident and there might be some short-term memory loss, but he'll soon be back to his responsibilities. As is often true of cloning narratives that imagine human duplicates beginning life as fully formed adults, a memory implant convinces each Sam of a singular history that is borrowed rather than directly experienced. Viewers slowly grasp that there was an original Sam who came to the station and received regular video messages from his loving wife, but several generations of clones have lived and died since this original's visit, each of which finds illusory hope in what are only endlessly replayed files. In fact, until Sam_1's accident (it eventually emerges that Sam_1 is really the fifth Sam), each clone has lived an unconscious repetition of the original's experiences and then sickened as he approached the expiration of his three-year life span. Voluntarily sealing himself within a coffin-like chamber that he expects to heal and return him to Earth, the clone is instead finished off with poisonous gas and stored away in the same hidden basement drawers from which his body, along with a fresh T-shirt, was elevated a few years earlier (fig. 9).

The interruption of this cycle that initiates the plot is Sam_1's rover crash and rescue by Sam_2, whom Gerty had tried to confine within the moon base. Portrayed by Sam Rockwell, this rapidly sickening clone and his fresh sibling are forced to acknowledge that their reality is not so straightforward as it seems and that neither is so unique as he thought. The film is remarkably unhurried in revealing the full significance of their duplicate appearances, allowing viewers to consider just how much evasion and procrastination human beings might undertake in order to avoid facing their own constructedness—the sheer fact that they did not have to be as they are. The reality *Moon* patiently builds is that these human beings have become not just unwilling but unconscious tools, living and breathing commodities immersed in a fictional reality that serves only corporate profits. This could have seemed easier to swallow if the clone had never become conscious of its exploitation, but the film eventually pulls away that blanket of temporary comfort, eventually showing us three different versions of Sam simultaneously (fig. 10). Like it or not, viewers are invited into the perspective of a human being who discovers he has been treated as anything but.

Sam_1 gets to the bottom of his situation when Gerty disobeys corporate instructions and provides the password that accesses video footage of the Sams before him, repeatedly living out the same sequences of desire and exhaustion and then sealing themselves into what they believe to be transportation home (fig. 11). Without relying on dialogue, Sam_1's stunned face recalls Kaph's recapitulation of his clone siblings' deaths in "Nine Lives," but it conveys an even more powerful shock. It is something like the moment when Jim Carrey's eponymous character in *The Truman Show* (1998) runs his boat aground on the enormous stage set that had constituted his life to that point. Addressing the man who had been directing his reality-show life since birth, Truman asks, "Was nothing real?" The same question lurks on the face of Sam_1, and what he and Sam_2 eventually face together is the discovery that they are individuals but also part of a larger organism. Most importantly, they remain capable of sacrificing for one another rather than the corporation that claims to own them. This is why Sam_2 lays the dying Sam_1 back in the crashed rover from which he earlier rescued him, and why, before dying, Sam$_1$ takes satisfaction in seeing Sam$_2$ shoot toward Earth in one of the pods used to transport the "helium-3" that the moon base mines. Before Sam$_2$'s departure, he risks his life as well, making a last-minute decision to step out of the pod long enough to destroy the communication jamming system that would keep Sam$_3$ and any other future clones from recognizing their scapegoating by humans back on Earth.

There are striking parallels here to Le Guin's "Nine Lives." Beyond the fact that both sets of clones are used as rather inglorious astronaut miners, they are harvesting resources needed by humans who have destroyed the ecological balances of their home worlds. Le Guin tells us little about the situation beyond her distant mining planet, Libra, but there is this understated revelation about Pugh's country of origin: "The United Kingdom had come through the Great Famines well, losing less than half its population" (456). The background of her story includes a cataclysmic collapse in Earth's system of resource production and distribution, and however it occurred, the work of the "Exploitation Corps" in harvesting other planets' minerals is tasked with overcoming it. The dependency is even more apparent in *Moon*, which begins with a fictional advertisement for the corporation manipulating the Sams, Lunar Industries. Viewers learn that the Earth had descended into a relatively dark, chaotic period due to insufficient energy, but the discovery and accessing of helium-3 on the moon renewed its former prosperity. Because of the work done by Sam and his clonal siblings, billions back on Earth are able to resume their former lives of profligate consumption, oblivious to their reliance on injustice elsewhere.

In both cases, this sacrificial labor is coded according to race and class, even if subtly. In "Nine Lives," we might overlook that Martin has an Argentine background and a "Hershey-bar-colored face" (455), while the clones all have "bronze skin" (454) and are named after an original with an Asian last name. Pugh is a white-skinned Brit, but in Le Guin's storyworld, that heritage does not protect him from suffering. The British apparently survived the roughest period of ecological collapse by hunkering down and executing black marketeers and hoarders; "where in richer lands most had died and a few had thriven, in Britain fewer died and none throve" (456). The registers of race and nationality are even more striking in Jones's film, both in the Korean language displayed throughout the station and in the awkward efforts of the two Korean executives' efforts to control Gerty from their conference room on Earth. Nowhere is their significance more evident, though, than in the reactions to Sam$_2$'s success in traversing the planetary boundary that had previously kept his sacrifice invisible. In the

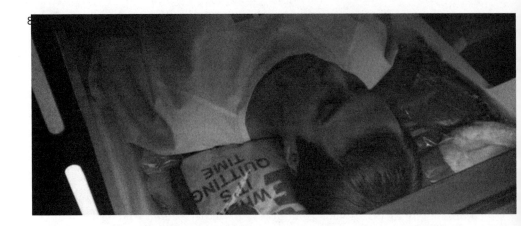

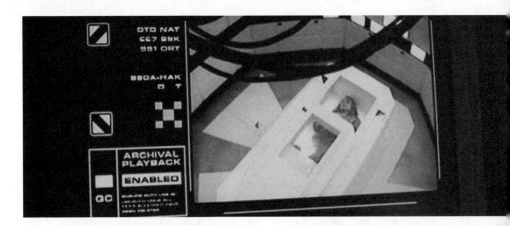

Fig. 9–11 The Sam clones are born, become aware, and die in *Moon* (2009).

film's final shot, as his pod descends into Earth's atmosphere, we hear overdubbed arguments on Earth about how the suddenly visible scapegoat should be treated. Loudest is the concluding radio show exclamation (voiced by Jones himself) that accuses Sam_2 of being "a wacko or an illegal immigrant—either way, lock him up!" In short, *Moon*'s clones have become an even greater threat than Le Guin's tenclone. Sam_2 is the exploited other who is risking everything in the hope of a new life, forcing average citizens to face their complicity in an abusive system they had previously been able to ignore.

In this sense, *Moon* demonstrates how the carbon-copy clone, tracked across four decades in this chapter, functions as the ultimate blank slate for scientism and objectivism. As imagined in genetic fantasy, the clone is a body shorn of or only loosely connected to a soul, one that can be manipulated for any ends desired by more powerful humans and their corporations. He can sacrifice himself to save others because, like the ideal automaton, he remains unconscious of himself as a unique source of need or value. She is simultaneously savior and scapegoat, much as René Girard's theory would lead us to expect, and yet unfortunately for the forces of corporatism, she refuses to stay locked in a state of unconscious passivity. As stories from Le Guin to Sargent to Butler to Jones illustrate, human beings do not belong in zoos, no matter how superficially comfortable the accommodations or how genetically similar the individuals. There might not be a specific material correlate for the soul within one's genome, but human beings cannot live without understanding themselves and others as sacred, at least in some pragmatic sense. Love depends on that fiction, whether expressed through shared grieving in "Nine Lives," insistence on sibling interdependence in *Cloned Lives*, readiness to trust the alien in *Xenogenesis*, or willing self-sacrifice in *Moon*. A healthy politics also requires narratives of the unique soul, as chapter 3 will bear out more fully in the realm of genetic realism. Just as genetic fantasy exposes the rationalizations by which many cultures privilege some sexual and gender identities above others and by which they abuse immigrants and their labors, so too does genetic realism. As I will show, however, that subgenre uses science not just metaphorically but also more literally. Genetic realism explores the means by which justice can be enabled or hindered not only through more plausible forms of bioengineering but also through character-driven, structurally complex accounts of individual and group interdependence. Even more than genetic fantasy, genetic realism is interested in the readiness of the exploited to stand up and fight back.

3.
The Cultural Determinism of Genetic Realism

More than a century before *Moon* used clones as analogies to an exploited working class, H. G. Wells's *The Island of Doctor Moreau* used human-animal hybrids to horrify readers at the prospect of unrestrained bioengineering. Before its violent climax, though, the novel allows its mad scientist a chapter to explain himself. Moreau says he is not insane but merely sees the world uniquely. Nearly alone on a remote island, he has been creating transgenic organisms that he calls humanized animals, and he is surprised only that other biologists have not already done so. The changes he has wrought are not just external; they transform "the chemical rhythm of the creature" (69). Among other modifications, he instills human language, assuming the animals will eventually be destroyed by humanity anyway, so he might as well experiment. In an often forgotten detail, he also declares his motives to be "religious," claiming that he follows in the steps of "this world's Maker." This is despite his complete disinterest in the animals' perspectives: Moreau no longer suffers the ethical hesitations caused by "sympathetic pain," having learned to assess the data before him without emotion. Pointing to his half-human puma, he announces, "The thing before you is no longer an animal, a fellow-creature, but a problem" (72). Like the "Negroid" (73) gorilla that was his first hybrid creation, it is an object awaiting manipulation.

Like most readers, the narrator, Prendick, is horrified by Moreau's exploitation of nature as well as his rationalizations. Initially afraid of the Beast Folk, Prendick soon realizes that their intermittent attacks are driven by an understandable terror. This is the same empathy for the abused creature that Mary Shelley's *Frankenstein* (1818) cultivated near the beginning of the nineteenth century, and it is no coincidence that the "poor, helpless, miserable wretch" (96) to whom she gives voice in her novel's middle chapters is echoed in Prendick's descriptions of the creatures in Wells's novel. They too are "poor brutes" (92) and "poor victims" (93), and our narrator loses both his fear of them and his confidence in life's meaningfulness. Using this generally noble if slow-witted foil to the novel's cruel biohacker, Wells builds on Shelley and anticipates Aldous Huxley in imagining a brave new world to come. Read today, some of these novels' experiments indeed seem more plausible than they did to their original audiences. Hybrid species now range from the cosmetic to the medically transformative, from cats that glow in the dark to mice that produce human antibodies.

That said, rather than debating the foresight of Shelley, Wells, or Huxley, I want to consider how these literary classics cultivated Western cultural fears of biologists "playing God." *Frankenstein*, *Doctor Moreau*, and *Brave New World* are key intertextual backgrounds against which to understand later innovations of genetic fantasy and then genetic realism. Chapter 2 showed how, in the second half of the twentieth century, new knowledge of DNA's structure and the increased plausibility of human cloning combined to produce many thrillers of genetic fantasy, on both page and screen. It also demonstrated how feminist science fiction authors repurposed the Carbon-Copy Clone Catastrophe masterplot in order to dignify the individual soul and its differences, whether encoded by race, class, gender, or sexual preference, and to drive home love's dependence on heterogeneity. What chapter 2 did not discuss extensively, however, was literary structure, partly because most of the texts engaged there are formally quite linear. From "Nine Lives" to *Cloned Lives*, and even from *Xenogenesis* to *Moon*, genetic fantasies tend to move chronologically from point A to point B, with only minor variations between story and plot. In these works, readers usually learn what happens in roughly the same order and in the same moment that primary characters do. It is not that genetic fantasies entirely lack flashbacks or foreshadowing; rather, their narrative and diegetic timeframes tend to match closely.

The situation is different for the more commonly circular, even spiraling genetic realism that started to appear at the turn of the twenty-first century. Reflecting a postmodern self-consciousness that grows even more apparent with the genetic metafiction of chapters 4 and 5, genetic realism is generally more committed to aligning with contemporary scientific possibilities than is genetic fantasy. To be clear, I am not suggesting that because a work is recent that it must be genetic realism rather than fantasy—*Moon* indicates otherwise—only that the widespread cultural awareness of genetics that cultivated genetic realism was relatively limited until the 1990s and 2000s. With the announced completion of the Human Genome Project, the general public grasped for the first time that it might be possible to examine the very core of a person's biological inheritance and come away with telling insights about their past, present, and future. This fostered new expectations about the meaning of genetics—hopes quite distinct from cloning fantasies' tendencies to rewrite earlier mythologies around the doppelgänger. Thanks to such doubles' prominence on screen, the possibilities of radical genetic engineering were on the radar for many Westerners well before the millennium, but few citizens had imagined that personal genomic testing might become an affordable reality within a decade or so. As we will see, genetic realism particularly revolves around the technology of personalized medicine, reflecting on the ways knowledge of the individual genome has become a form of high-tech divination. Rather than relying on superheroes, aliens, and other versions of the deus ex machina, these stories narrativize the gene in ordinary human lives, often spanning large generational and cultural divides in the process.

This chapter's reverse chronological trip through genetic realism begins on the edge of fantasy with the BBC America television series *Orphan Black* and then works backward toward the turn of the millennium. The first two texts I will consider, *Orphan Black* and Kazuo Ishiguro's novel *Never Let Me Go*, are relatively realist adaptations of the cloning tale. Like their limitlessly fantastic cousins in chapter 2, they use extensive theological references and play with the idea that clones appear to be mirror images of one another, but these narratives employ alternate-history approaches toward quite different ends. The TV series tweaks history by imagining that scientists secretly defied restrictions enacted by their colleagues in the mid-1970s and successfully created two lines of human clones, with at least the female members all born in 1984. Ishiguro's novel is less specific in dating its events, using a prefatory page to tell the reader only that the subsequent events are set in "England, late 1990s." Given that Kathy H. is thirty-one at the novel's opening, this puts her birth in the late 1960s and places scientists' work on the first human clones sometime in the 1950s—well before even the most ambitious post–World War II timelines might have imagined.

Aside from these temporal novums, *Orphan Black* and especially *Never Let Me Go* adhere relatively closely to the boundaries of realism, even if the television series also utilizes more conventions of satire and thriller. More than previous cloning tales, these narratives develop powerful arguments for the idea that in the most profound ways, *we are all clones* (beginning with the fact that we share 99.9 percent of our DNA with all other humans). These stories' protagonists function not as carbon copies or stereotypes of particular cultural groups but as dynamic characters with whose transformations viewers and readers are invited to identify. The Canadian show uses a combination of religious language, phallic imagery, unique parent-child relationships, and historical references to eugenics to challenge genetic determinism and corporatism and to affirm the depths of human uniqueness and the unconditional nature of love. Likewise, Ishiguro's novel relies on subtle intertextual references and other clues to align the reader with the novel's clones and to expose how genetic determinism actually functions as a form of *cultural* determinism.

With this paradox established, the chapter's third section reaches back to two of the earliest major novels of genetic realism, Zadie Smith's *White Teeth* and Jeffrey Eugenides's *Middlesex*. They aptly represent the interest among major literary artists in the Human Genome Project's implications for human meaning making. Ian McEwan's *Saturday* (2005) is another especially vivid example, with its neurosurgeon protagonist concluding somberly of a Huntington's sufferer whose life he has temporarily saved, "This is his dim, fixed fate, to have one tiny slip, an error of repetition in the codes of his being, in his genotype, the modern variant of a soul, and he must unravel" (279).[1] What especially fascinates me about Smith's and Eugenides's novels, though, is how they weave concern about overreactions to new genetic knowledge into an equally profound interest in how transnational movements shape individual and generational identity. In some ways anticipating the increased visibility of the epigenetics movement, these novels study the combined impact of twentieth-century shifts in immigration and biotechnology. Taken together, they suggest how the metaphors of genetic biology cultivate not only newly hybrid senses of nationality and ethnicity but also increasingly complex sexualities and gender identities. They show genetic realism, attesting that absolute purity is a dangerous myth and that no matter how great the technological advancement, the past is never past. Ironic repetitions are inevitable, but examined closely, they also contain the seeds of difference. For genetic realism, instead of hiding the subtle distinctions that compose each soul, we must find meaning within them. Smith's and Eugenides's novels envision genes and their environments dynamically interacting; far from eliminating personal agency, they stress its enduring power.

Genetic Realism on the Edge of Fantasy

This book's introduction explored how another Canadian television show from the mid-2000s, *ReGenesis*, took pains to represent pandemics and bioterrorism with unusual biological precision and to flesh out the lives of scientists as fully as possible. Also shot in Toronto,[2] *Orphan Black* took up *ReGenesis*'s mantle and now stands as one of the most thorough explications of the epigenetic tension between genes and environment ever to appear on screen or page. Beyond the quality of its writing, acting, and postproduction, the foundation of *Orphan Black*'s success is its alignment of feminist, queer, and ultimately postsecular critiques against a too-easy biotechnological corporatism while still maintaining a deep-seated enthusiasm about the positive potential of genetic research and new medical technologies. Embodying an intertextual consciousness that has become a predominant trait of genetic fiction, this TV serial builds on not only Shelley, Wells, and Huxley but also lesser-known, more recent novels like Sargent's *Cloned Lives*. In the process, it demonstrates how genetic influence is both very real and yet only part of the picture. Perhaps most strikingly, it asks how love can be described by but still exceed biology, suggesting that this depends on defying religious fundamentalism's and global capitalism's mutual complicity in human objectification.

The show's alternate-history premise is that a combination of corporate and government interests began secret experimentation with reproductive human cloning soon after the 1975 Asilomar Conference on Recombinant DNA, long before Dolly the sheep's birth announcement in 1997 and just as bioethicists, government watchdogs, and most scientists were beginning to think it possible. The resulting children are now adults, but not all are aware of their origins. In the first two seasons, viewers are invited to identify with three clones in particular: Sarah, initially a negligent mother prone to disappear for a year at a time and to make ends meet selling drugs, a habit patiently resisted by Felix, her gay stepbrother; Alison, an obsessively organized suburban soccer mom with two adopted children and an overweight, always-snooping husband, Donnie; and Cosima, whose freshly minted PhD in genetics allows her a unique perspective on the activities of the show's fictional Dyad Corporation, even as her dreadlocks and lesbian self-discovery land her in a relationship with a woman who is revealed to be one of its top scientists, Delphine. Then there is Helena, the Ukrainian avenging angel hell-bent on murdering her "sestras." Helena has been brainwashed by a religious cult, the Proletheans, that raised her to believe that her clone sisters are the demonic copies of her original source material, and much of the early plot turns on her decisions about whom to believe. As it turns out, Alison and Cosima are aware of the threat, having already been in contact with other clones like Beth Childs, the police detective whose suicide Sarah witnesses in the pilot's opening scene and whose identity she assumes in an attempt to access the woman's bank account. To say that "complications ensue" vastly understates *Orphan Black*'s intricacies, and only determined viewers can stay cognizant that all these characters are played by a single shape-shifting actress, Tatiana Maslany. This is to say nothing of the male clones who emerge in the show's third season or of additional developments in seasons four and five.

Season two is particularly evocative in its exploration of the relationships between literal and figurative children and parents, the latter of whom sometimes suffer from divine pretensions. I examine it here as a microcosm of the entire show's interest in the dialogue between creators and creatures, a twenty-first-century expansion on the relationships between Frankenstein and his creature and between Moreau and the Beast Folk.[3] One of two highly paternalistic figures in the show's first two seasons, Dr. Leekie is a corporate geneticist whose "mad scientist" role is intimated by his first name, Aldous. This technoenthusiast has developed his own sense of morality, and it should surprise no one who has read the preceding chapters that his TED talk–style sales pitches are steeped in transcendent rhetoric. In season one, he recruits Cosima to a lab at the Dyad Corporation, at first condescending to her as a graduate student but soon realizing that she is not intimidated by his fame and that her dissertation on "the epigenetic influence on clone cells" has prepared her to grasp the significance of his efforts toward "patenting transgenic embryonic stem cells"—another allusion to Huxley's novel and its hybrid-species experiments. It is not coincidental that Cosima first encounters Leekie as he is promoting "Neolution," a cult-like transhumanist movement. Offering his listeners the possibility of replacing their current visual ability with infrared, X-ray, and ultraviolet capacities, he enthuses, "Plato would have thought we were gods." In season two, he waxes similarly poetic before potential investors at a fundraising party for Dyad: "To combine is to create; to engineer, divine," he declaims. This is humanity pursuing divinity not with humility but via high-tech mimicry, a pulse-pounding vision of the future that defies the inevitability of death and views genetics and other cutting-edge sciences as tools for elevating the species into a mystical invulnerability.

If Leekie's language exploits religion for technocapitalist purposes, the show's other major cult uses biotechnology to serve religious ends. The Proletheans are a group of seemingly low-tech traditionalists living on what appears to be a self-sustaining communal farm. However, their exceedingly modest dress code and decorum mask a heavy investment in the tools of artificial insemination and genetic modification. As Henrik Johanssen explains of the effort to use his sperm, Helena's eggs, and as many "brood mare" women as possible to expand his clan, "Man's work is God's work, as long as you do it in his name." His public prayer is equally revealing; he informs God, "We are your instruments in the war for creation." But Henrik does not just rely on apocalyptic biblical allusions and militant, paternalistic rhetoric. Beyond the extremist stereotype, he also possesses some attractive characteristics. Like Leekie, Henrik is awed by genetic biology, embracing its findings as revelations rather than threats to his faith, even if he is similarly overconfident of his ability to control life. Played by Peter Outerbridge, the same actor who helped create the more sympathetic researcher David Sandstrom in *ReGenesis*, this sexist is blind in his convictions. Yet we also see him leading a children's story time with genuine charm, amusingly adapting Shelley's novel to create the same happy ending he expects to foster in real life. "His creation pursued him with a terrible vengeance, because the doctor had never shown his creation any love," Henrik tells his enrapt young audience. "And so when they finally came face to face, they sat down, and they had a great big bowl of iceberg cream!"

Unfortunately for the storyteller, his own ending cannot be sugarcoated, and ultimately, the audience is not sorry. Henrik never learns one of *Orphan Black*'s and indeed genetic fiction's foundational lessons: love is antithetical to use. The unquestioning patriarchy of Prolethean culture may allow him effectively to take Helena as a second wife, remove her eggs, inseminate them, and then place the embryos in her womb and in that of his daughter; however, it is no coincidence that the show portrays him adapting the same tools to impregnate women as he does cattle—they are no less experimental beasts than the humanized animals in Wells's novel. Appropriately, when Helena finally escapes her bedroom prison and overcomes Henrik (with his daughter's help), he finds himself strapped into the same stirrups he used to access his patients' wombs. Tied in place, he panics as he slowly grasps the clone's intentions. Marshaling the same farm husbandry implements he had used on her, Helena gleefully asks how far his interest in human-animal hybridity goes: "Would you like horse baby? Cow baby?" The last we hear of the Prolethean leader is a terrified scream as she shoves the lengthy insemination device through the upper reaches of his anal canal. Helena's triumph is as appalling as it is just, and it represents the rawest form of *Orphan Black*'s feminist rejection of the patriarchal technoreligious manipulation that Wells imagined a century earlier.

Beyond its shock value, two further elements of this scene deserve attention. First, however brutal Helena's actions, they are motivated by a defense of her "babies," as she calls them. While less conscious of social expectations than the other female clones, Helena embodies a childlike innocence that is matched only by her fierce instinct to protect the vulnerable. At the end of the scene featuring Henrik's *Frankenstein* adaptation, for instance, she observes one of the Prolethean women disciplining a distracted child with needless cruelty. Pinioning her against a wall, Helena informs the woman that she will be "gutted like a fish" if she does something similar again. Second, the phallic shape of Helena's vengeance against Henrik is not just a clever device for transfixing the audience. By utilizing his own artificial insemination stick, she turns his penetrative power back on him, creating the most painful of ouroboros images. There is nothing pretty about the outcome, but its reversal of men's violence against women is riveting. A woman raised by a religious cult to believe that she and her sisters are "abominations"—a commonly decontextualized biblical translation routinely leveled at LGBTQ individuals and sprinkled across the series, starting with the fourth episode of season one—rejects their ideology, turns their violence on them, and departs to defend her true family.[4] It is no mistake that the scene's denouement lingers on Helena's face as she looks back on the burning Prolethean farmhouse. Like *Frankenstein*'s creature departing the burning cottage where he had learned to read but was ultimately rejected, Helena is thoroughly disillusioned with her early mentors.

This is far from the only moment in which *Orphan Black* redeploys a phallic signifier in order to illustrate the nonutilitarian nature of authentic love and its sexual expression. Not all these scenes are so serious: when Alison's husband proves impotent with a jackhammer, for instance, the results are comical. Failing to break the concrete in their garage under which they will (repeatedly) bury the accidentally murdered Leekie, Donnie hands her the gas-powered battering ram, scoffing at the notion that she might do better. Alison breaks through the surface immediately and turns to him with a smirk, and their eventual success in completing the unconventional interment proves an aphrodisiac. *Orphan Black*'s references to phallic power often anticipate violence, though. One of the most emotionally intense sequences in the show's history comes in season two's fourth episode, when Sarah slips into the condo of Dyad's new leader, her clone sister Rachel, who was raised by the corporation after the disappearance of her early childhood parents, Ethan and Susan Duncan. Eventually caught by one of Dyad's hired guns, Sarah is forced into a glass-enclosed shower and handcuffed to the overhead fixture. After sharpening his razor, the henchman begins an excruciatingly slow process of cutting her throat. The show's avenging angel answers her prayers, however: Helena bangs into the apartment, still wearing the exceedingly modest wedding dress supplied by the Proletheans, and promptly dispatches Rachel's thug. But this is hardly good news to Sarah, as she now shrinks from what she fears will be a new assailant, given that she had shot Helena the last time they met. The camera lingers over Helena's hip-high, upturned knife blade as she approaches, but instead of finishing the male torturer's violence, Helena shocks her sister into convulsive tears, falling onto Sarah's breast like an exhausted child seeking a mother's embrace (fig. 12).[5]

Why is this scene so moving? Not just because of the way Sarah escapes the razor wielded by Rachel's minion but also because Helena declines to turn the knife on her sister. If the point were not sharp enough, it is repeated in the next episode, when Sarah convinces Helena to put down a sniper rifle rather than giving Rachel what she too might seem to deserve. Looking through the glass wall of an adjacent skyscraper, Sarah and Helena see their lingerie-clad sister straddling Paul, who replaces the henchman dispatched by Helena in the previous episode. Significantly, he is not allowed to enjoy the sexual services he provides, earning a slap when he reaches for Rachel. This again illustrates how the show reverses but also reaches beyond a form of sexual objectification usually applied to women: Rachel commands him not to kiss her and to be still as she pleasures herself but remains entirely unaware that Helena's crosshairs rest on her skull. Sarah steps into her sister's line of sight, determined not to let Helena shoot, and the sniper's initial response again demonstrates *Orphan Black*'s stress on love's distinction from use. "You only want to use me," Helena accuses Sarah. But her clone sister proves convincing, seemingly discovering the truth of her words even as she utters them: "No, that's not true. You saved my life. You're my sister. I thought I killed you. I couldn't tell anybody what I lost." Reenacting the shower scene of the previous episode, Helena surrenders a different pointed weapon, hoping once again what experience has taught her to doubt—that love might not be delusory. There is nothing weak, passive, or sentimental about this choice. On the contrary, *Orphan Black* reaches beyond the thriller's stereotypical

Fig. 12 *Orphan Black* uses queer imagery between clones Helena and Sarah to drive home a vision of familial love. *Orphan Black* season two, episode four (2014).

boundaries to demonstrate that an even greater power can imbue acts of mercy rather than of violence.

Taken together, scenes like these represent *Orphan Black*'s feminist and often queerly inflected rejection of the corporate utilitarian power driving genetic determinism, whether it is being used to fuel religious fundamentalism or scientistic reductionism. It is not enough for Helena merely to take revenge, whether on Sarah or on Rachel in these scenes or on Siobhan in season three: what she wants is genuine acceptance. Only hope in the possibility of loving and being loved is capable of making a trained killer trust a woman who had previously stabbed and shot her, and it is one of many places in which the show demonstrates a sober optimism about individual agency. Not only does Helena grow immensely in her capacity to believe in others—though not without serious relapses—but Sarah becomes far more responsible, Alison far less self-centered, and Cosima far more willing to accept others' help. In these ways, *Orphan Black* builds significantly on the unique clone characters in the 1976 novel that we explored in chapter 2, Pat Sargent's *Cloned Lives*. The TV series insists not only that environment can make radically different characters of virtually the same genetic material but also that individuals can learn to make profoundly different choices from those to which they are predisposed, even when a corporation claims ownership of their DNA.

Orphan Black's intertextual relationship with *Cloned Lives* is tightened further by the fact that Sarah's little girl, to whom Cosima owes her life and Helena her smile, is named Kira. Like the Kira who serves as the only female main character in Sargent's novel, *Orphan Black*'s central child character is a beacon of peace around which much of the violence paradoxically revolves. Cal makes her an origami butterfly, but she interprets it as an angel, sending it with her mother as a form of protection ultimately embodied by Helena. Kira's significance grows as we come to understand Rachel not just as a corporate slave but as a clone who has been the ward of a research institution since the Duncans were taken from her in the middle of her childhood. While most often appearing cold as steel, this Rachel also weeps for a lack of children, much like her biblical namesake.[6] In fact, Rachel's deepest anger at her "father" (while not genetically related, Ethan is the scientist who supervised her birth and early years) stems from his insertion of a genetic sequence that rendered most of the clones incapable of sexual reproduction. In one intercut sequence, viewers watch her imagine destroying Leekie's office, even as she actually maintains her composure and attempts to reverse her fate. So envious is Rachel of Sarah's mysterious exceptionality that she kidnaps Kira, placing her in a pink-and-white stage-set bedroom at Dyad in the delusory hope of taking over as a maternal figure. At this point at least, she lacks the self-awareness to recognize that she is attempting to repeat the worst part of her own upbringing. She does not begin to grasp the distinction between using Kira and loving her. By contrast, when Cosima and Sarah arrange Kira's escape, the show emphasizes the lengths to which a truly unconditional love will go, even if it also relies on a great deal of luck. Sarah gets only a single shot with her improvised fire-extinguisher cannon (yet another phallic metaphor), and even this daring rescue depends on the complicity of corporate forces that exceed Rachel's authority.

The contrast between Rachel's attempt to manipulate Kira and Ethan's attempt to raise Rachel is also instructive. Absent from Rachel's life for decades, Ethan stirs memories that leave her weeping before old home movies. We see how easily she dispatches with Dr. Leekie once his disloyalty becomes clear, acknowledging only that having received his "nurture" means she will give him a chance to survive, rather than having him eliminated immediately. Yet she is far more attached to her previous father figure. When Ethan succeeds in committing suicide with a poisoned bag of tea before he can be forced to reveal genetic sequences that would let Rachel control her sisters more fully, she is simultaneously enraged and heartbroken. She screams in frustration, "You can't leave me again!" Ethan is the anti-Frankenstein who loved his creature from the beginning—"My poor, poor Rachel," he gasps with his dying breaths—but it seems not to have been enough. Like his daughter, he ends up a pawn in a larger corporate game. Just as Leekie is dispatched when his use is exhausted, Ethan was elbowed out of the way so that the corporation could program Rachel with an abstract, heartless worship of biotechnology. He wanted Rachel to live freely and know that she was loved, but he failed. At the same time, he recognizes that her choices played a role, lamenting with his final breaths, "I'm afraid you don't deserve me anymore."

Ethan did accomplish something before dying, though. Unlike Rachel's obsession with Kira, Ethan's care for Sarah's daughter is authentic. In their brief acquaintance, he becomes a kind of surrogate grandfather, and before leaving for his final conversations with Rachel, Ethan gives Kira a book, *The Island of Doctor Moreau*. The volume turns out to contain not just the text of the novel but an extensive system of coded marginalia and scribbled notes that ensure his knowledge of the clones' unique genetic sequences will not be lost. The passage from chapter 14 that he reads aloud to Kira is telling: in recognizing the danger in thinking of "an animal, a fellow creature, [as] a problem," Ethan resists Moreau's objectifying rhetoric and aligns himself with the megalomaniac's servant, Montgomery.[7] This is suggested by season two's finale, when just before his suicide, Ethan watches an old video of himself reading to Rachel as a child. The fragment we hear ("And looking round sprang to my feet with a cry of horror") is again from *Doctor Moreau*, this time a climactic scene in chapter 19. Moreau is dead at that point, and Montgomery is trying to strand Prendick on the island by burning his boats and his hut. Similarly, Ethan commits suicide in hopes of stranding Rachel without the genetic knowledge she would need to take control of her sisters.

Why does Ethan take such an extreme step? Perhaps because he can see just how wrong the project has gone in the decades since the clones were born. What began with his love for science as a means of acquiring knowledge has produced a tool for rendering scientism an unquestionable ideology. The male clones that appear in season three might seem even more obvious illustrations of this objectification, given their development by the military as a crack team of covert operatives, but their corporate owners would use the female clones no less dispassionately. Nonetheless, nature has resisted complete human control, producing Sarah as a "wild type" who is somehow fertile. As Ethan explains to Cosima, the project's original scientists inserted codes into the clones' genome that were supposed to "degrade the endometrium, prevent the ovarian follicles from maturing." This is what causes Cosima's follicular lymphoma, which spreads to her lungs and threatens to kill her just like it did two earlier female clones. Appropriately, then, in sacrificing himself in order to protect the women from their would-be overlords, Ethan arranges for Cosima to take his place as a surrogate parent for Kira. Already close, Cosima and Kira are drawn together further in Sarah's absence. Most tellingly, when Kira awakens Cosima the morning after her escape, Cosima initially mistakes the little girl for her lover, Delphine. This leads to another story time in which Cosima reads Kira a popular Canadian picture book about an exceedingly loyal dog, and the show completes its transference of Ethan's scientific and parental roles. Kira hands Cosima the special copy of *Doctor Moreau* she received from Ethan, and the famously nightmarish tale of bioengineering gone wrong provides new hope for saving the young scientist from her cancer.

By joining humility to science's endless thirst for knowledge, characters like Ethan and Cosima raise *Orphan Black* above many screen treatments of biotechnology. Refusing to paint scientists in entirely light or dark tones, it is neither naïvely enthusiastic nor unduly condemnatory about genetics. On one hand, it leaves little doubt that new knowledge and tools are capable of inspiring and enabling fanatic cults that yield only self-destruction and judgment, whether they are enthusiastically garish like Leekie's Neolution or purportedly minimalistic like Henrik's Proletheans. On the other hand, the show weighs both varieties of megalomania against the integrity and idealism represented by Ethan, Cosima, her colleague Scott (perhaps the show's most lovable nerd), and Delphine (once she finally abandons Dyad). Indeed, Cosima's character was directly inspired by the show's scientific consultant, Cosima Herter, whose job was to stay current on the most relevant work in genetics and let that knowledge inform the show's screenwriting, set design, acting, directing, and editing processes. She helped *Orphan Black* achieve unusual heights of scientific detail, as when Scott explains that a sequence in the clones' DNA is "anomalous for cytochrome C" or another character prepares blood for recovering stem cells via "apheresis." Yet *Orphan Black* not only gets most of its technical details right but insightfully assesses relationships between biology and feminism. Herter's PhD work at the University of Minnesota was in the history of science, technology, and medicine, which starts to explain the whole *Orphan Black* team's unique success in engaging not only specific genetic details but also the cultural contexts in which biology operates.[8] This creates a much larger consciousness of social, philosophical, and theological implications than is typical for a television show, one evident as much in its episode titles (taken from such writers as Charles Darwin, Francis Bacon, and Donna Haraway) as in its deepest existential reflections (like Cosima's season three riff on *Hamlet* while holding a brain she just removed from its skull).

The most poignant historical connection made by season two provides one last insight into how genetic realism shapes audience attitudes toward biotechnology. Alerted to the fact that Ethan is still alive, Sarah tracks him to an old church that houses the records of an institution called "Cold River," which Helena describes as "the place of screams." Posing as a graduate student interested for research purposes, Sarah soon finds herself holding physical evidence of the facility's abuses, which are thinly veiled reflections of practices at the Cold Spring Harbor Laboratory, home to America's infamous Eugenics Record Office. As Sarah files through century-old materials, viewers glimpse a "Cold River Breeding Study" volume; a photo captioned "Most Perfect Baby, 1908"; a page on "Spastic Idiocy"; and many other references to "cretins," "feebles," "sex perverts," and other associations with criminality.[9] All these have direct correlates with actual documents from the real Eugenics Record Office; one of the images Sarah examines (fig. 13), for example, is a minimally photoshopped negative image of one of the real institution's photos, "Massachusetts Department of Mental Diseases Exhibits Pictures of 50 Criminal Brains" (fig. 14). In other words, the alternate history that drives this serial television—the secret cloning of Sarah and her siblings in the 1970s—is inextricable from early twentieth-century eugenics. The cloning project created by Ethan and Susan Duncan and then assimilated by Dr. Leekie, the Dyad Institute, the military, and the cabal revealed as "Topside" is fictional heir to one of American history's most disturbing actual biocultural projects. The US effort to build "Fitter Families" at county fairs and a "healthier," more homogenous population—the same propaganda that would inspire Hitler's eugenics program in Nazi Germany—is the foundation of the institute that brings Sarah and her siblings into the world. Indeed, in season five, *Orphan Black* reveals that Neolution has been behind behind a contemporary program with similar aims.

Given this association with what we must unapologetically label "evil," what might be most remarkable about *Orphan Black* is that it still refuses to descend into the knee-jerk reactions against genetic biology that are typical of

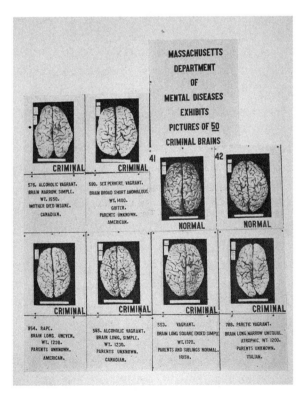

Figs. 13–14 Actual material from the Eugenics Record Office is incorporated into *Orphan Black,* season two, episode six (2014). "Massachusetts Department of Mental Diseases Exhibits Pictures of 50 Criminal Brains," Image 567, Cold Spring Harbor Laboratory Archives, 1921, http://www.eugenicsarchive.org/eugenics/.

many popular narratives. Certainly, there are potent attacks on the potential for genetic knowledge and modification to become tools wielded only by the powerful, and it is no mistake that the history of such exploitation is hidden in a religious institution's basement, much like the space where Helena lives temporarily in season one. However, even as Sarah shares her findings with her sisters, Cosima demonstrates that it is possible to acknowledge fully the risk of biotechnological abuses and nonetheless remain deeply devoted to her science's positive potential. Unlike Leekie and Henrik, she has no need to perform a false certainty or gloss over her field's many areas of incomplete knowledge. "Science is what scientists do," she acknowledges. "We're just poking at things with sticks." This is not reason for despair but greater investment, trusting that the "sticks" will become more precise and efficient with time, and perhaps even that the operative metaphor can become less phallic. Cosima insists that her discipline can yield knowledge that will turn into expressions of love, not just fear. Indeed, it is Kira's willingness to donate stem cells from her baby teeth and then from hip bone marrow that allows Cosima to keep fighting off her cancer. She is living testimony that genetic medicine is capable of much good, even as she is also the apparent product of bioethical failures. *Orphan Black* indicates that biotechnology's actual manifestations can be messy, ambiguous, and self-contradictory, but it rejects absolutist calls to abandon such tools. Without question, an obsession with technotranscendence can bypass all manner of safeguards in efforts to create possibilities deemed "inevitable," but there remains much positive potential for new genetic knowledge and applications to be employed responsibly and to save and improve lives.

Just as *Orphan Black* explores exciting new avenues of epigenetics, then, it refuses to remove biotechnology from its historic contexts. The questions that this series and other works of genetic realism ask about the nature of the soul and its capacity to love must remain open inquiries. For that reason, Cosima tells Delphine that if she is going to love her, she must "love all of us," not pit her and her sisters against one another. This advice extends to the audience: learning to make the most of genetic biology's new capacities means recognizing the imbrication of the individual soul and her communities, whether such persons fit normative definitions easily or just "need a bit more help" than average—as Michael Bérubé says of people with Down syndrome in *Life as Jamie Knows It*, his 2016 update of his son's story (169). It means accepting that there are more than obvious causes and effects in the world, that stories are not just linear arrows shooting forward to change the future but reverberating spirals that remake our understandings of history, which in the process enable new futures. It is fitting that at the end of season two, Cosima shows Sarah her spiraling, shell-like tattoo of the golden ratio as she encourages her sister, observing that it repeats itself in every molecule of their DNA. The hymn "Nearer My God to Thee" plays over the season's closing montage as the show's various subplots draw toward moments of temporary closure (in the marriage of Henrik's daughter) that immediately promise to reopen (as we discover that her new husband is one of the male clones featured in season three). The result is a sense not only that genes will never stop shaping identity but also that their expression is always dependent on environment, particularly that created by one's closest relationships. For *Orphan Black*, being near the divine might mean many things, but to be whole means defying those who use knowledge to enslave and risking everything for those who are most vulnerable.

We Are All Clones

Before there was *Orphan Black*, there was *Never Let Me Go*. Published in 2005 and adapted as a film in 2010, Kazuo Ishiguro's novel is the first text in which the cloning fantasy mutated into genetic realism.[10] The innovation here is quite different from that of *Orphan Black*, as Ishiguro only minimally engages the biological specifics of human cloning. At the same time, this is not genetic fantasy like the Le Guin short story, Sargent and Butler novels, and Jones film explored in chapter 2. Technotranscendence is nowhere to be found, neither as ideal nor as threat. Furthermore, the uniqueness of the individual human clone is never in question for the reader—only for the introspective young characters and their society. Ishiguro takes readers into the psychology of a clone character more fully than any novel before or since, creating an alternate history whose only significant departure from realism is to imagine that Britain began human cloning for the sake of organ donations even earlier than *Orphan Black* posits. The result might be twenty-first-century literature's most intense exploration to date of Girard's scapegoat mechanism, a novel that joins *Orphan Black* in asking readers to recognize that in a profound sense, we are all clones. For *Never Let Me Go*, biotech utopians and would-be eugenicists are redundant: Western culture *already* teaches its citizens to replicate, objectify, and police themselves, all without questioning their societies' true sources of power or concepts of justice. The novel also offers a stunning rebuke to the assumption that education necessarily serves democracy, or at least to the notion that democracies necessarily value true equality of opportunity. These social criticisms do not make Ishiguro uninterested in the actual realities of genetic biology but indicate only that his realism has different aims from *Orphan Black*'s. Without concerning itself with the specifics of genetic biology, the novel focuses on illuminating the individual psychology and cultural anthropology necessary for any society to foster love when there are radical and systemic differences in opportunity, whether caused by biotechnology or other forces.[11]

As chapter 2 demonstrated, genetic fantasies about clones often seem to concern a different species, with characters fundamentally different from the reader. These are beings not created by God but by human hubris, we are told, the result of biotech's mastery of life to an extent Mary Shelley never contemplated. The fleshly materials her scientist retrieves from graveyards and then sparks into crude new life-forms are now the province of sophisticated laboratories and sterile birthing vats. The products are controllable, unquestioning beings who are purportedly free from the great existential anxieties that beset the rest of us. "Who am I?" is rendered irrelevant, or at least easily answerable. The publics of such cloning tales expect their victims to respond, "I am merely a repetition of an efficient pattern with a clear purpose. I go places too dangerous for regular people, I provide services beneath your dignity, I sacrifice so you may thrive." In fact, such an "I" is not a person at all but only an artificial approximation. The assumed measuring stick for the clones of genetic fantasy is economic use-value, not the love idealized by *Orphan Black*. Such beings are attractive to imagine because they can reproduce more quickly and with fewer flaws, can work longer without complaint, can be replaced without moral qualms when their useful lives expire. Or at least, that is the narrative used to establish many cloning narratives—until the creatures rebel, discover some semblance of personhood, and attempt to claim lives more like those of their creators.[12]

In this work of genetic realism, however, Ishiguro constructs a nearly three-hundred-page novel in which rebellion never occurs to his living health insurance policies. This eliminates suspense; the novel goes out of its way to forecast plot developments, preempting any possibility of thrills or spectacle. Instead, *Never Let Me Go* lulls readers into the same sense of apathy besetting its characters, "students" we only gradually realize are human clones. Reared in the most insular of private-school circumstances, they will one day become organ donors for the individuals back in regular society who supplied their genomes.

We follow Ruth, Tommy, and our narrator, Kathy H., as they graduate from their idyllic school, Hailsham; enjoy a two-year transitional period full of sexual dalliance (without need for prophylactics, due to their genetically assured sterility and culturally decided isolation from disease); serve as "carers" who facilitate older clones' donation and "completion" processes; and then become donors themselves. Through its combination of memoir and bildungsroman, *Never Let Me Go* slowly insinuates that not only are we all already clones but in a real way, genetic determinism is better understood as *cultural* determinism.

Girard's thinking is invaluable for illuminating the psychological process Ishiguro explores. As the French theorist argues, "Mimeticism is the original source of all man's troubles, desires, and rivalries, his tragic and grotesque misunderstandings, the source of all disorder and therefore equally of all order through the mediation of scapegoats" (165). Like Ishiguro's clones, most humans learn to want things because others want them. On the basis of only semiplausible threats, political and religious rituals ostracize a select few so as to create a mirage of warm inclusion for everyone else. These walls are erected at a safe distance, so that "everyone participates in the destruction of the anathema but no one enters into direct physical contact with him. No one risks contamination. The group alone is responsible. Individuals share the same degree of innocence and responsibility" (177). Ishiguro's clones, in other words, are the people to whom society denies the status of persons even as it simultaneously depends on them to stitch clothing, harvest food, and clean bedrooms for unconscionably low wages. In fact, they are everyone who accepts a hierarchy that claims some people are more equal than others.

The irony of Ishiguro's clone narrative is that it demonstrates how easily *nurture* can turn people into clones. It imagines a group of literal clones and then shows the same scapegoat mechanism operating within this group as outside it (e.g., in Tommy's treatment by his classmates). What is often presented as an alternative class of being instead emerges as ordinary humanity hidden under a blanket of superficial uniformity. A novel that nominally explores the most cutting-edge of future innovations turns out to be far more interested in historical and ongoing exploitation of the underprivileged. Tellingly, *Never Let Me Go* features not a single glimpse of the technology that enables cloning, describing only the downtrodden rehabilitation spaces in which clones attempt to recover from up to three or four "donations" before their inevitable "completions." The novel is also unique among cloning tales in never featuring a clone meeting its match or its original. In one case, the characters search out such a "possible," but the revelation that the doubling was illusory only emphasizes the point.¹³ In short, the novel aims not to imagine what kind of unique entity an actual bioengineered clone might be but to hold up a mirror to its readers.

Each of the three main characters offers a unique opportunity to grasp the psychological consequences of objectification. When Tommy feels isolated, abandoned, and betrayed, his outrage comes out in his "raving, flinging his limbs about, at the sky, at the wind, at the nearest fence post" (10), and the novel's readers recognize their own experiences of isolation, abandonment, and betrayal. Ruth is desperate to feel the unique object of another's love and affection, to be "show[n] favouritism" (57) by Miss Geraldine with the pretended gift of a cheap pencil case, and Kathy H. acknowledges, "Didn't we all dream from time to time about one guardian or other bending the rules and doing something special for us? A spontaneous hug, a secret letter, a gift?" (60). Then there is the narrator herself, who tries to make sense of things by telling her story even as readers marvel at her incapacity to see beyond her own experience and recognize her captivity within an unfair system. But how much more aware are most people? If these clones fail to rebel against the culture that determines their futures, Ishiguro's novel asks, are "regular" individuals' backbones any stronger?

Beyond demonstrating the illusory nature of the "artificial" individual's difference, then, *Never Let Me Go* demonstrates that genetic determinism itself is actually a form of cultural determinism, one rendered all the more effective by its biological mask. The clones of Ishiguro's novel never assure themselves of their identity with a personal genome test, but they regulate themselves far more effectively than anyone outside of their system might. Hailsham is the most compassionate of prisons and thus all the more effective in generating the complacency and submission it requires. Were there guards, guns, and barbed wire, the clones might recognize the possibility of escape. Instead, these teenagers are freely allowed to leave Hailsham, spending a roughly two-year transitional period in halfway homes like the Cottages before volunteering to enter facilities where they begin "donating" body parts. It is the fact that they have such free rein across the countryside that makes them so incapable of imagining any alternate destiny. Even the myth of a three-year "deferral" period for couples in love contributes to this sense of inevitability, as it gives what little hope they have a specific but harmless avenue, one they can dream about but never fulfill. Miss Lucy, the one guardian with an active conscience, laments, "Your lives are set out for you. You'll become adults, then before you're old, before you're even middle-aged, you'll start to donate your vital organs. That's what each of you was created to do. You're not like the actors you watch on your videos, you're not even like me. You were brought into this world for a purpose, and your futures, all of them, have been decided" (81). The problem is that the clones believe her—but, Ishiguro suggests, so do we.

As blind as the clones of *Never Let Me Go* remain to the possibility of rebellion, Ishiguro designs his narrator's testimony so that readers can recognize more common forms of oppression. In the trick played on Tommy when he gashes his elbow, for instance, we might notice the same dualistic notion of the clone body as empty container that appears in Hollywood blockbusters. His classmates warn that if he isn't careful to keep his arm straight or if he loses his temper again, his skin might fully "unzip" and his elbow will "pop out" (87). This is not just a single joke but a whole mythology that expresses the clones' unacknowledged anxiety about their future donations. "The idea was that when the time came, you'd be able just to unzip a bit of yourself, a kidney or something would slide out, and you'd hand it over," Kathy H. explains (88). There are also cracks about unzipping and dumping body parts into a hungry child's bowl, and what they reveal is an entire culture's effort to separate the physical from the emotional and the spiritual.[14] The organs aren't really "you" but just something you can hand over, like cash to pay your taxes.

There are also disturbing hints that the "teachers" who have the most reason to recognize the full humanity of their "students" will still take advantage of their society's perverse medical safety net. In order to rationalize this choice, they too imagine a clear line between souls and bodies. When Kathy H. and Tommy finally meet Miss Emily again at the novel's conclusion, Madame rolls her colleague out of the darkness in a wheelchair. She has hidden herself in the shadows for the first part of their conversation, and when she appears, she is preoccupied by the need to sell a "beautiful object," her bedside cabinet, without it being damaged in transit. Perhaps used to store medicines, the cabinet is a synecdoche for the clones themselves: both are storage devices needed only temporarily, objects that must be protected not for their own sakes but so that their resale value is not diminished. The most revealing detail of the conversation is what Madame says immediately upon appearing before them, the significance of which Kathy H. and Tommy entirely overlook. "I've not been well recently," she admits, "but *I'm hoping this contraption isn't a permanent fixture*" (257, emphasis added). Implicitly, even one of the most "caring" of the guardians, a woman who dedicated her life to achieving greater quality of life and more humane conditions for future organ donors, plans to trade the life of a clone for her personal convenience.

If this seems too subtle, consider how Miss Emily's readiness to sacrifice the life of a former student fits the entire novel's metaphorical pattern. Ishiguro inserts repeated references to physical objects with use-value like that of the clones. For instance, the two older clones with whom Kathy H., Ruth, and Tommy travel to Norfolk are excited about buying packs of Christmas cards at a Woolworth's, a department store that epitomizes the values of redundancy and repetition. One of the senior clones concedes, "You end up with a lot of cards the same, but you can put your own illustrations on them. You know, personalise them" (157). This is the same hunger to assert individual difference conveyed by "Nine Lives," *Cloned Lives*, *Xenogenesis*, and other post–World War II cloning fantasies. Here, though, the symbolic significance of the relevant objects is far more subtly conceived. Ishiguro makes another such move when Kathy H. comes across a desk lamp in a shop window that matches the ones where she is narrating the action years later: "Here in my bedsit I've got these four desklamps, each a different colour, but all the same design—they have these ribbed necks you can bend whichever way you want" (208). There is a constant undercurrent, a subconscious awareness, that there is something wrong here, that these objects carry additional weight, that there are human beings in this novel who also "bend whichever way you want." But the connection is never rendered so directly that we cannot overlook it. Like Miss Emily's bedroom cabinet, the clones only gradually come into focus as items that can be used and then discarded as long as one takes care to prevent unnecessary damage.

At this section's outset, I noted distinctions between *Never Let Me Go* and *Orphan Black*, particularly the novel's lower interest in biomedical verisimilitude. However, there are also deep parallels, like the works' shared reliance on Shelley's *Frankenstein*. In the BBC America show's case, the reference is to the disparity between creators who are willing to objectify their creations, like Dr. Leekie and Henrik, and those who remain committed to them, like Ethan. Ishiguro's novel employs a similar *Frankenstein* reference to expose just how false the guardians' performed sympathy is. When Madame speaks with Kathy H. and Tommy at the novel's conclusion, she refers to them three different times as "poor creatures," the same sentiment originally expressed in Shelley's novel, echoed in Wells's, and picked up most recently by *Orphan Black*. In case the connection is not yet convincing, Madame also explains how Hailsham's failure was decided by a rogue geneticist who went beyond human cloning to engineer new traits that might prove superior to those of the average population. This renders the novel's public far too conscious of the potential personhood of its organ sources, and it illustrates how very distinct genetic innovations can be conflated in the court of public opinion, a phenomenon I call "biotech slippage."[15] The key detail Miss Emily lets slip is that the scientist "carried on his work in a remote part of Scotland, where I suppose he thought he'd attract less attention." This is the same setting to which Victor Frankenstein travels in order to expand his grisly labors and create a second creature. In both Shelley's and Ishiguro's novels, one offense against nature leads to another. Miss Emily explains the necessity of cloning organ donors, regardless of the ethics: "There was no way to reverse the process. How can you ask a world that has come to regard cancer as curable, how can you ask such a world to put away that cure, to go back to the dark days?" (263). Yet as readers captivated by Kathy H.'s narration, we *are* that "world," which suggests that the system was not inevitable within Ishiguro's storyworld any more than it is for our world. We might even find ourselves wondering if Frankenstein's open rejection of his creation, or at least his eventual refusal to repeat his mistake, is not preferable to Madame and Miss Emily's repressed shuddering upon each encounter with the clones. Their superficial attempts to render palatable a horrific system of exploitation are motivated by the worst forms of condescension and pity.

More fully than any previous cloning tale, and in ways surprisingly parallel to *Orphan Black*, then, *Never Let Me Go* provides poignant testimony that being branded a clone is a fear all human beings must face. Ruth is hardly alone in copying mannerisms from television shows, and Kathy H. might as well be an ordinary parent telling a child that some things aren't worth copying, even as she does the same thing herself. For our narrator, the fear is that her genetic stock comes not from the "dynamic, go-ahead types" (144) that Ruth admires but from "*trash*. Junkies, prostitutes, winos, tramps. Convicts, maybe, just so long as they aren't psychos" (166, emphasis in original). What this demonstrates, counterintuitively, is that in Ishiguro's novel, the clones are the most eager to find their unique identities, but they are also the greatest genetic determinists of all. They are desperate "to forget for whole stretches of time who we really were" precisely because they are so fully convinced that their identities are decided by biological inheritance (142). This is why they call the people who might be their genomic originals "possibles"—not so much because they are possible matches but because they represent what the clones take to be the possible heights of their own achievements. The possibles might be other people completely, raised in radically different environments and making entirely distinct individual choices, but they are the images of what might have been for the clones. Occasionally, the more insightful clones recognize the wrongheadedness of this assumption; Tommy objects, for instance, that he doesn't see how it matters to find one's possible or "what difference it makes to anything," and Kathy H. adds, "It's daft to assume you'll have the same sort of life as your model" (165). But like the smartest "normal" humans, it is difficult for them to hold onto this awareness. Only pages later, we find Kathy H. searching through pornographic magazines for her likeness, indicating once more that the so-called genetic determinism driving her search is very much the result of social pressures, not any knowledge of actual biology.

Like *Orphan Black*, then, Ishiguro's genetic realism takes readers deeply into the minds of its clones, characters that it suggests already exist. Sometimes we play the role of the guardians, teaching their charges to gloss over untimely deaths and injustice with euphemisms about "not making it" or "completing." Other times we're the clones themselves, accepting predetermined fates with only the faintest resistance. When Tommy asks Madame and Miss Emily at the end of the novel, "Why train us, encourage us, make us produce all of that [art]? If we're just going to give donations anyway, then die, why all those lessons?," the questioner and the respondents momentarily blur into a single identity (259). Why write novels, read them, or write about them if there is no escaping death, even if it is decades rather than a few years away? Ishiguro is too much an artist to attempt any final, absolute answer, but *Never Let Me Go* provides a moving beginning to one, a lesson that often seeps through the greatest works of literature. Echoing Ecclesiastes, Ishiguro's novel stresses that transience is intrinsic to beauty. Vulnerability is a precondition for wisdom; real stories cannot abide perfection. It is no mistake that at the beginning of Ishiguro's tale, the character who spends his childhood raging most violently at the world's cruelty—at his scapegoating at the hands of children who are also scapegoats—is called "the idiot!" (7). We might also note the precise language of Miss Emily's telling explanation to Tommy about Hailsham students and their "dream of being able to *defer*" (261, emphasis in original). Whether nodding to Fyodor Dostoevsky or Langston Hughes, Ishiguro signals that his utopia-turned-dystopia is about more than the dangers of a purportedly unbounded transhumanist future. Like much genetic fiction, it is about the opportunities awaiting human beings who have been misled by their biological traits to accept culturally imposed limitations. It is about people who face injustice now and can wait no longer for others to stand against it.[16]

Ishiguro's novel was the first to offer a full-scale adaptation of the cloning fantasy to the strictures of realism, but earlier works of genetic realism also pointed obliquely to cloning's significance. One of the most scientifically detailed turn-of-the-millennium works of genetic realism was Gwyneth Jones's *Life* (2004), a globetrotting assemblage of insights about genetics, artificial intelligence, and the enduring sexism and homophobia of early twenty-first-century culture. With just a few sentences from her female scientist protagonist, Jones offers a multifaceted view of the relationship between individual and community. A year before Ishiguro's novel would emphasize the distinctiveness of the individual soul, even one whose DNA matches that of another person, Jones's narrator observes, "You couldn't work long in human genetics without becoming conscious of how extraordinarily alike we all are. Practically identical, interchangeable units. (So why all this fuss about cloning?) Whereas, on the other hand, individuals change within themselves every year, every month, every day under different pressures and in different circumstances. Choose one human being, arbitrarily, who suits you well enough. Stay with him, and you'll see the whole human race go by" (186). If *Never Let Me Go* displays just how deeply individual circumstances and choices can define a person, *Life* gestures to a converse truth. From a technical point of view (whereby only 0.1–0.2 percent of human DNA differs between individuals), no matter how superficially diverse *Homo sapiens* might be, its members have far more in common than we sometimes pretend.

Nonetheless, Jones's novel ultimately shares the emphasis on valuing difference that is found in Ishiguro's. As understood in *Life*, genetics shows that "there's no such thing as normal" (88). Most critically, "sex is now something we can take apart and change around" (114). The novel gets profoundly right the ambiguity, unpredictability, and nonmechanical nature of genetic inheritance and mutations:

> At school, Anna had been thrilled by the certainty of the DNA process. It was such a trick, so satisfying and neat, the two complementary strands of bases, unzipping, acting as templates for replication: safe as the ticking of a watch. At school they let you think perfect replication was the norm—with the occasional derailment, so that new species could be born. When you got closer, you realized what happened was totally different. The process was weak, not strong. The strings of bases were continually

being repaired, continually evading repair, the patterns snagging and dropping stiches, so it was amazing you woke up every morning and found that a rabbit stayed a rabbit and a rose a rose. Coherent change emerged, mysteriously, like new music, out of the constant noise of miscopying: sections stuffed in backwards, upside down, in totally the wrong place. . . . How could this flux of meaningless chemical glitches drive the engine of something so powerful as evolution? (53)

As with *Orphan Black*, *Never Let Me Go*, and the two works I consider next, Jones's tale illuminates how regularly apparent accident imbues the genetic patterns that we are measuring and replicating with such increasing precision. This reality grows especially visible as *Life* challenges binary thinking about sex and gender.[17] As a news article in the novel says of maleness and femaleness, "*So much of our philosophy, of our humanity itself, rests on this most vital opposition. I seriously wonder if we can be human without it*" (334). The same question asked three decades earlier by Le Guin's *The Left Hand of Darkness*, this probing of easy oppositions is even more central to Jeffrey Eugenides's *Middlesex* and Zadie Smith's *White Teeth*.

Eugenides told me that he did not read Smith's novel before completing his own, and I do not doubt him. Rather, this is why I find so significant the uncanny parallels between his 2002 multigenerational tale about Greek immigrants in Detroit and Smith's 2000 multigenerational tale about Bengali immigrants in London. Both feature World War II scenes in which main characters manage to escape the actual fighting, both include extended narration from a state of preconception consciousness, and both imagine characters joining Islamic fringe organizations. Most critically, Smith's and Eugenides's novels demonstrate genetic realism's dedication to a balancing act that both affirms science and challenges scientism.[18] Both works are informed by considerable genetic detail, but they do not attempt bioethics in fictional form. Instead, their convergent evolution indicates growing cultural concern at the turn of the millennium with fatalism, purist rhetoric, and control obsessions associated with biotechnology and biomedicine. In the face of scientism's performed certainty, these novels affirm the scientific values of curiosity and precision but also the scientific realities of uncertainty, mutation, and hybridity. They envision new possibilities not just for embracing ethnic and sexual others but also for reimagining fiction's purposes in the genome age.

It is possible to put aside the science and read either novel primarily as a cross-generational tale, an immigration narrative, or in *Middlesex*'s case, an examination of intersexuality. These approaches have dominated literary criticism to this point, and they make it clear that Smith's and Eugenides's works constantly challenge the myth of purity, whether expressed in terms of race, nationality, sex, or gender.[19] As Alsana Iqbal tells her husband, Samad, in *White Teeth* many years after their immigration to the United Kingdom, "You go back and back and back and it's still easier to find the correct Hoover bag than to find one pure person, one pure faith, on the globe. Do you think anybody is English? Really English? It's a fairy tale!" (196). Her son Millat illustrates this perfectly, infuriating his father by being "neither one thing nor the other, this or that, Muslim or Christian, Englishman or Bengali; he lived for the in between, he lived up to his middle name, *Zulfikar*, the clashing of two swords" (291). On the other hand, his twin, Magid, becomes obsessed with purity as both a Muslim and a young man. As Alsana complains after he returns from an involuntary year spent in Bangladesh, "His teeth, he brushes them six times a day. His underwear, he irons them" (350). This amusingly echoes the narrator's description of Hortense Bowden, Jehovah's Witness par excellence, who reveals that she always wears two pairs of panties so that on the Lord's return, she can "whip off the one closest to her and replace it with the outer pair, so that Jesus would find her fresh and odorless and ready for heaven" (328). Unfortunately, such discomfort with the body expresses a deeper prejudice: her fixation on purity leads her to condemn interracial marriage, concluding that "black and white never come to no good. De Lord Jesus never meant us to mix it up" (318). *White Teeth* repeatedly skewers this brand of self-righteousness.

Middlesex's opposition to purity fetishes is no less urgent. The grandmother in Eugenides's novel is Desdemona Stephanides, who as a young woman lived in a small Greek village near Smyrna with her brother, Lefty. As the Turks send the Greek army fleeing in 1922, she and her brother slowly accept the fact that they are in love and embark on the kind of incestuous romance that *Cloned Lives* rejects. Following their comedic scheme to meet as purported strangers and marry on a refugee boat to America, Desdemona's heritage continues to shape her choices, especially via her mother's axiom that "to have good silk, you have to be pure" (22). However, when she attempts to pass along this wisdom to the Nation of Islam women whom she teaches to make silk in Detroit, her new boss insists that it not be applied beyond textiles. When it comes to ethnicity, the job requires her to admit, "Everybody mixed. Turks, Greeks, same same" (145). Her granddaughter, Calliope, who decides near the novel's conclusion that she can function better in society as a he, Cal, has far less trouble with blended natures.[20] Compared with the popular girls at their upper-class private school, our narrator notes, "'Ethnic' girls we were called, but then who wasn't, when you got right down to it?" (298). Cal/lie already grasps that purity is its own costume, one adults attempt to purchase by sending their kids to British-sounding schools with strict uniform codes, since "in America, England is where you go to wash yourself of ethnicity" (337). Her parents even go the extra mile by buying a house with architecture that imitates Frank Lloyd Wright, using open floor plans in its "attempt to rediscover pure origins" (273).

Yet neither *White Teeth* nor *Middlesex* stops with rejecting the illusion of complete, uncomplicated purity. Many of their greatest insights lie in deconstructing superficial commitments to the apparent alternative, hybridity, which can too easily become a means of masking colonialism and assimilation. Smith's compositional artistry is evident in her brief introductions of characters that later become more important, even if by that time readers do not always remember them.[21] Marcus and Joyce Chalfen, the geneticist and his gardening wife who become central to the novel's second half, first appear as "an aging hippie couple both dressed in pseudo-Indian garb" at the same school board meeting where Samad Iqbal makes a pain of himself by criticizing the secularism of the "Harvest Festival" (110). The Chalfens join Samad's doomed resistance in a moment that might seem peripheral, but it sheds light on their later motives for inviting Samad's son Millat as well as Irie Bowden into their home. These study sessions aspire to collect the ethnic other, just as in the school board scene, their attire claims cultural currency through a veneer of international sophistication. Similarly, their patience for Millat's rebellions lets them imagine themselves as open-minded, even if the real purpose is to allow Marcus greater proximity to the Iqbal son he is truly interested in mentoring, Magid. Unfortunately, that relationship too features ulterior motives, as Marcus wants to ensure that there will be a scientific protégé to honor his influence.

In *White Teeth*, in other words, hybrid identities are an unavoidable aspect of the human condition, yet the novel criticizes superficial enthusiasm for difference. On one hand, self-anointed prophets like Samad proclaim, "We are divided people" (149), "split people" (150). Hybridity creates establishments like O'Connell's Poolroom, which is entirely free from Irish cuisine or billiards but features "an Irish flag and a map of the Arab Emirates knotted together" (153). It makes a youth gang into a "cultural mongrel" of Nation Brothers, Raggastanis, and Pakistanis, which together follow an ethos that variously cites Allah, Bruce Lee, and Public Enemy. Yet the question begged by these admixtures is whether unlike things are simply being jammed into marriages of convenience that will break apart once sufficient pressure is applied. Or are they true fusions, transformations of each component so that they find new life in relationship? Smith's novel sends its youthful protagonists to the Glenard Oak School, which was built in two stages and represents 67 different faiths and 123 different languages, but this doesn't create wholeness or unity any more than does the assimilationist propaganda of the Ford English School in *Middlesex*. In Eugenides's novel, when an immigrant dressed in Balkan costume descends into the melting pot constructed on stage for the school's graduation ceremonies and then emerges alongside immigrants from Syria, Italy, and Poland, all dressed in matching wool business suits—with "thirty-two teeth brushed in the American manner" and underarms "liberally sprinkled with American deodorant" (104)—the result is not the integration of cultures but their suppression. This assimilative process is also represented by the church Lefty visits in Greece before he becomes an immigrant. There we see him praying for deliverance from attraction to his sister in a former mosque, with "Koranic inscriptions plastered over to provide a fresh canvas for the Christian saints that are, even now, being painted on the interior" (31). Finally, the adaptation of *The Minotaur* that Lefty and Desdemona watch in Detroit might be the most damning critique of false hybridity in either novel. Just as it features "a bull's head attached to a human body," the play is an Americanized spectacle only loosely connected to its Greek origins, a chimera that doesn't take (108).

All these insights about purity and hybridity are accessible without attention to Eugenides's or Smith's treatments of science. Read as genetic realism, though, *White Teeth*'s and *Middlesex*'s curious affinity and distinctions become even more revealing. These novels reject false performances of national, racial, and sexual purity particularly through their attention to rising expectations for genetic testing and modification.[22] Smith and Eugenides reimagine immigration as a low-tech form of genetic engineering, one that mixes genes and cultural environments and yields new ways of inhabiting both race and gender. The novels indicate that a major problem with humanity is its susceptibility to endless repetition, so each work proposes its own solution for getting off the hamster wheel of samsara and reconceiving biological inheritance. Smith's narrative emphasizes the role of accident, while Eugenides's story shifts the burden further toward individual agency. Taken together, they demonstrate genetic realism's capacity to lay bare the potential of biotechnology, both criticizing its rhetorical overreaches and acknowledging its potential to transform medicine.

The success of their balancing acts depends heavily on readers' awareness of narrative shape. By implication, effectively adapting to a rapidly accelerating knowledge of DNA's double helix depends on approaching human stories as spiral, not just linear or circular.

At first, Smith's novel seems to present the fatalism driving relationships between the generations as infinitely circular. Soon after meeting Archie Jones, we learn that he was once an Olympic cyclist incapable of completing a lap in either more or less than 62.8 seconds. This lack of variation is just as great a burden for his friend Samad Iqbal, whose family epitomizes the way "immigrants have always been particularly prone to repetition—it's something to do with that experience of moving from West to East or East to West or from island to island. Even when you arrive, you're still going back and forth; your children are going round and round" (135–36). Samad's explanation is that Western lifestyles are overly controlled, that "these first descendants of the great ocean-crossing experiment . . . live in big plastic bubbles of our own creation, their lives all mapped out for them" (182). The problem, in other words, is that there is insufficient room for accident, *White Teeth*'s most frequently invoked solution to cultural entrapment. While youth like Marcus Chalfen's son Joshua and Samad's son Millat embrace determinism and the belief that "we live in circles," the novel's wisest characters recognize that some things are beyond explanation (419). From Archie's encountering Clara on the commune staircase in the novel's opening pages, to Millat's falling and breaking his nose immediately upon receiving word of his brother's having done the same thing, to the convergence of destinies at the concluding press conference to announce FutureMouse, *White Teeth* is a raucous testament to the roles of mishap and choice in shaping human lives. This is the atmosphere within which the novel initially defines genetic engineering's significance: it appears to characters like Samad as one more tool by which Western culture would eliminate chance and control the future.

Of course, the novel's central geneticist, Marcus Chalfen, is not worried in the least about increased predictability. He welcomes biotechnological control over uncertainty. One of the clearest insights into his motives comes via his wife, Joyce, a popular gardening columnist who blithely characterizes his hybridization work as that of a twenty-first-century Moreau:

> Her husband didn't just make money, he didn't just make things, or sell things that other people had made, he *created* beings. He went to the edges of his God's imagination and made mice Yahweh could not conceive of: mice with rabbit genes, mice with

webbed feet (or so Joyce imagined, she didn't ask), mice who year after year expressed more and more eloquently Marcus's designs: from the hit-or-miss process of selective breeding, to the chimeric fusion of embryos, and then the rapid developments that lay beyond Joyce's ken and in Marcus's future—DNA microinjection, retrovirus-mediated transgenesis (for which he came within an inch of the Nobel, 1987), embryonic stem cell-mediated gene transfer—all processes by which Marcus manipulated ova, regulated the over- or under-expression of a gene, planting instructions and imperatives in the germ line to be realized in physical characteristics. Creating mice whose bodies did exactly what Marcus told them. And always with humanity in mind—a cure for cancer, cerebral palsy, Parkinson's—always with the firm belief in the *perfectability* of all life, in the possibility of making it more efficient, more logical (for illness was, to Marcus, nothing more than bad logic on the part of the genome, just as capitalism was nothing more than bad logic on the part of the social animal), more effective, more *Chalfenist* in the way it proceeded. (259–60, emphases in original)

Like *Orphan Black*'s Dr. Leekie and even the guardians in *Never Let Me Go* who believe they are giving "students" their futures, Chalfen is a control freak armed with pipettes and petri dishes. Another worshipper at the throne of technotranscendence, he pits himself against literary theorists and philosophers like "those strange French men who think truth is a function of language, or that history is interpretive and science metaphorical." We are back to the ideology of objectivism here: "Truth was truth to a Chalfen. And genius was genius." The act of interpretation is nowhere to be found, nor is ambiguity, nuance, or deviation from expectations. Instead, as his wife explains it to herself, "*Marcus created beings*. And Joyce was his wife, industrious in creating smaller versions of Marcus" (260, emphasis in original). Yet readers also see her naïveté. Smith's novel repudiates its geneticist's determinism, even going so far as to link it to the carbon-copy version of cloning that we saw operating in genetic fantasy. Chalfenism is inherently, cheerfully sexist: like the clones of *Never Let Me Go*, Joyce seems unaware of her exploitation, but through Irie and other characters, *White Teeth* joins earlier genetic fiction in exposing the predominantly male voices behind a scientism that would use women to gain a perverse, illusory sense of eternal life.

The physician at the end of *Middlesex* who tests Callie's genome and evaluates her gender does not come off much better. Dr. Luce's office door plaque informs patients that he specializes in sexual disorders and gender identity, and he is indeed careful to distinguish between these categories during the two weeks he interviews his prize patient. However, readers who catch his other name, John Money, cannot be surprised when Dr. Luce proves a hypocrite, even if the novel never quite calls him out directly.[23] He tries to convince Milton and Tessie that Callie must be surgically altered in order to conform to her sex of rearing and presumed identification. When pressed, he admits that it's more complicated than this, that both genetics and culture play roles: "It's not a matter of sheer genetics. Neither is it a matter of purely environmental factors. Genes and environment come together at a critical moment" (464). However, he tells his colleagues, as one of them summarizes, "Chromosomal status has been completely overridden by rearing" (421). This is a far more extreme claim than the nuanced position Milton and Tessie force him to acknowledge, and it is also what he presents in the draft of his journal article about the case that Callie happens to see. The result is his patient's choosing life as a male and fleeing her family rather than accepting surgical alteration. Cal/lie's parents succumb to the same "enticement" that led Desdemona and Lefty to pretend they were not siblings: "No one would ever know" (429). The difference, of course, is that genetic testing can now reveal what history might otherwise repress.

One might expect *Middlesex* to treat the unique secondary sex characteristics of its main character with a rhetoric of "accident" similar to that found in *White Teeth*. After all, only about one in two thousand babies are born with intersex genitalia. And there are indeed some moments in which the novel seems to blame happenstance. As the elderly Lefty endures increasingly frequent ministrokes, for instance, we find him "confronting the possibility that consciousness was a biological accident. Though he'd never been religious, he realized now that he's always believed in the soul, in a force of personality that survived death. But as his mind continued to waver, to short-circuit, he finally arrived at the cold-eyed conclusion, so at odds with his youthful cheerfulness, that the brain was just an organ like any other and that when it failed he would be no more" (263). This prophecy proves accurate. Like Cosima's Hamlet-esque meditation on the brain-soul relationship in *Orphan Black*, Cal seems to embrace a language of sheer accident in describing Lefty's death: "Blood pooled in his brain for the last time, washing even the last fragments of his self away" (269). Likewise, the narrator's brother responds to the threat of Vietnam by "taking chemical trips of his own . . . trying to escape what he had dimly perceived while wrapped in an afghan: the possibility that not only his draft number was decided by lottery, but that everything was" (317–18). The option is on the table: perhaps we are to understand inheriting 5-alpha-reductase pseudohermaphroditism as merely another expression of meaningless chance, the infinite jest with which the universe mocks individual human fates.

However, Eugenides places a weight on agency that goes beyond his and Smith's attention to accident and that has key structural implications. Here we find a provocative inversion of Richard Dawkins's famous approach in *The Selfish Gene*, wherein human beings and other organisms function as puppets for the gene's purposes. In *Middlesex*, by contrast, genes are dependent on personal and familial narrative. Indeed the most distinctive aspect of Eugenides's work is its unprecedented narrativization of the gene as a culturally and personally determined actant. The novel's hinge lies at the midpoint between book two and book three, at the end of a chapter entitled "Ex Ovo Omnia" ("Everything Comes Out of an Egg" 198). Readers struck by the strange name given to the narrator's brother, Chapter Eleven, can count their way through the unnumbered table of contents and discover that Eugenides calls attention to this interpretive plot summary even before the novel's first sentence. The end of the novel's eleventh chapter returns to the narrator's conception and explains that within Cal's mother, "a billion sperm swim upstream," and that they "carry not only instructions about eye color, height, nose shape, enzyme production, microphage resistance, but a story, too." This tale began "two hundred and fifty years ago, when the biology gods, for their own amusement, monkeyed with a gene on a baby's fifth chromosome" (210). Several generations later, the effects reverberate in the immigrant journey of young Lefty and Desdemona and continue in the lives of Milton and Tessie, just as they will reappear in Callie's unique journey to becoming Cal: "Hitching a ride the gene descended a mountain and left a village behind. It got trapped in a burning city and escaped, speaking bad French. Crossing the ocean, it faked a romance, circled a ship's deck, and made love in a lifeboat. It had its braids cut off. It took a train to Detroit and moved into a house on Hurlbut; it consulted dream books and opened an underground speakeasy; it got a job at Temple No. 1. . . . And then the gene moved on again, into new bodies" (210).

Middlesex takes the cultural narrative about genetics that popular rhetoric makes readers expect—that the gene is the all-determining source of an individual's fate—and upends it. The novel's midpoint summation facetiously treats a "gene" as if it were capable of independent agency, but the novel as a whole shows how individual characters responding to historical pressures actually make these choices. As a result, Eugenides's personified gene emerges less like absolute fate and more like the fire in Smyrna that "dodges," "leaps," "tightrope-walks," and is "shocked by its good fortune" (56), or the fog in San Francisco that "creeps," "shrouds," "walks," and "clouded . . . minds" (469). In *Middlesex*, the gene is synonymous with tortuous personal decisions, not a means of glossing over those choices with determinist rhetoric. As the narrator summarizes near the novel's conclusion, "Free will is making a comeback. Biology gives you a brain. Life turns it into a mind" (479).

Life does this—or, examined more distantly, history does. *Middlesex* and *White Teeth* engage several key cultural moments that reveal their spiral integrations of continuity and change. In Smith's novel, accident might be a major theme, but the structure is clearly set up to make readers feel the tension between Archie Jones's situations in Bulgaria in 1945 and in London in 1992. As a young man, Archie proved incapable of making a choice, and a captured Nazi war criminal, a renowned eugenicist referred to as Dr. Sick, escaped his grasp. Almost a half century later, he is finally ready to act, and the novel's conclusion has him save the man for a second time. The radical difference is that this time, he makes a choice, whereas in 1945, he was mired in indecision. He had known then that Dr. Sick, or Dr. Marc-Pierre Perret, had been part of the Nazis' "sterilization program, and later the euthanasia policy" (90). And his only friend in the battalion, Samad Iqbal, had informed Archie that Dr. Perret's so-called science was profane: "Choosing who shall be born and who shall not—breeding people as if they were so many chickens, destroying them if the specifications are not correct. He wants to control, to dictate the future. He wants a race of men, a race of indestructible men, that will survive the last days of this earth. But it cannot be done in a laboratory. It must be done, it can only be done, with faith! Only Allah saves!" (100). Still, this simplistic appeal to the opposition of the religious and the secular proved insufficient for Archie to act. Throughout most of the novel, readers are led to believe with Samad that after Archie traded his poker winnings for his fellow soldiers' silence and the rights to their prisoner, he had taken Dr. Sick into the darkness and shot him. Instead, the novel's final pages reveal that Archie could not bring himself to pull the trigger and that the eugenicist escaped, only to reappear in the novel hundreds of pages later (initially unrecognized by many readers) as Marcus Chalfen's scientific mentor, Dr. Perret.

This uncanny recurrence is of an entirely different variety from Archie's 62.8-second laps. We now need Peter Brooks's definition in *Reading for the Plot*: "An event gains meaning by its repetition, which is both the recall of an earlier moment and a *variation* of it: the concept of repetition hovers ambiguously between the idea of reproduction and that of *change*, forward and backward movement" (99–100, emphasis added). *White Teeth*'s narrative is not circular so much as spiral: there is both circularity and linearity, repetition and difference. The older Archie again saves Dr. Perret from a bullet, but whereas this was previously the result of inaction, here it represents active courage. At the same time, the threats in the two scenes are closely linked. In the first case, the danger is Nazism's eugenic project of social engineering; in the second, it is scientism's hubris in deciding the life span of a genetically modified mouse (and implicitly, imagining similar control of human beings). While Hitler's minions were performing heinous experiments on people and Dr. Chalfen and his mentor are, for the time being, only manipulating mice, both scenes warn against a technotranscendent lust to control life.

For *White Teeth*, to control the gene is to control the plot—but use too firm a hand, and the overdetermined result will be endless circularity and predictability. This is the problem with the ideology of objectivism: it seduces the most brilliant minds into performances of certainty that they should not actually feel. Just as Dr. Luce in *Middlesex* knows better than his hyperbole about nurture overriding genes, Dr. Chalfen in *White Teeth* is caught up in a hope to "eliminate the random" and therefore "rule the world" (282–83). As detailed in his press release, the FutureMouse project promises to "slow the progress of disease, control the process of aging, and eliminate genetic defect," inaugurating "a new phase in human history, where we are not victims of the random but instead directors and arbitrators of our own fate" (357). He could be Madame in *Never Let Me Go* looking past life in a wheelchair, Dr. Leekie selling investors on the Dyad Corporation in *Orphan Black*, or the characters of *Cloned Lives* looking on starships as disability-free prayers to the universe. Yet Marcus does not really believe his own rhetoric, or at least he has not fully thought it through. He knows human destiny involves more than genetics; he argues in a private letter to Magid that he and his brother actually prove the argument against genetic determinism. The problem is not that Dr. Chalfen is evil in any straightforward way but that he prioritizes his public image as a scientist over what he knows to be demonstrably true. Unfortunately, there is often a gap between the actual data and the kind of assuredness that academia and the stock market reward most dramatically, and sometimes the short-term payoff for glossing over inconsistencies proves too tempting.

Just as the episode titles for season four of *Orphan Black* are taken from the writings of Donna Haraway, Smith told me that she was reading Haraway while composing her novel. *White Teeth*'s conclusion is very much a fictional riff on the feminist scholar's 1996 book about, among other things, DuPont's trademarked OncoMouse. The mouse might only appear near the novel's conclusion, but throughout it, Smith joins Haraway in challenging the ideology of an incontrovertible, hierarchical objectivism. To pretend one can anticipate every influence on an individual life, even one so tightly controlled as that of Chalfen's mouse, is delusory. And there is a further objection made by the young Indian woman reading Chalfen's book at the airport: "Actually creating an animal just so it can die—it's like being God!"[24] She goes on, "I mean personally I'm a Hindu, yeah? I'm not religious or nothing, but you know, I believe in the sanctity of life, yeah? And these people, like, *program* the mouse, plot its every move, yeah, when it's going to have kids, when it's going to die. It's just *unnatural*" (346, emphasis in original). The brilliance of the scene is that Smith unveils both the woman's insights and her unfounded fears. Marcus is indeed driven by an addiction to control, even by biotechnology's pseudodivine manipulation of the received natural order. Yet the novel also observes how rapidly the young woman conflates one bioethical issue with another, reading as eugenic a legitimate scientific effort to uncover patterns in disease susceptibility and increase public health. While evincing biotech slippage, she also takes the scientist's narrative to endorse an entirely circular version of history, one that would eliminate accident so as to control not just a mouse's story but all life. Instead, the novel implies, there must be a more spiral approach, a way of acknowledging historical and genetic influence that unlocks the cycle of absolute predictability.

Without that mind-set, eugenics makes a perverse kind of sense. Like *White Teeth* as well as *Orphan Black*, Eugenides's novel directly connects contemporary medical and biotechnological capacities to a much longer history of cultural determinism and ethnic discrimination. In Smith's case, the link is to Nazi Germany; for *Orphan Black*, the references are to the American eugenics movement. With *Middlesex*, the ties are to a whole host of early and mid-twentieth-century forces of paternalism, from British colonialism to American assembly-line industrialism and denials of African American civil rights. As early as Lefty and Desdemona's escape from their homeland's pillaging, a British ship captain abandons thousands of Greek refugees to die at the hands of the Turks, declaring there is "nothing you can do with these people" and regretting only the good cigars that will be lost with the city (52). Eugenides's real target, though, is his own nation's contempt for "undesirables," portraying Ellis Island as less a welcome center than another hoop to jump through, even for those who prove acceptable (73). Much as Cal searches for a new identity by submitting his long hair to a barber decades later, his grandmother is shorn of her immigrant braids by the YWCA ladies beneath the Statue of Liberty. And this is only a preview of the ways in which she and her husband are expected to conform in the Land of the Free.

Paramount to my argument is how *Middlesex* joins *Orphan Black* in representing biotechnology in continuity with the narrative of American corporatism and industrialism. The process by which immigrants like Lefty and Desdemona are taught to lose their cultural identities in the roaring twenties shares a great deal with the formula by which African American civil rights were resisted throughout the twentieth century and by which contemporary Americans intrigued by genetic testing might be led to devalue embodied uniqueness in favor of homogenous ideals. Cal's grandparents become part of a lower-class labor force devoted to manufacturing matching vehicles for richer Americans, a role that requires conformity to their company's expectations, just like the products they build. Unfortunately, Lefty's night school diligence and other efforts to learn the role of a clean, dutiful citizen prove insufficient because he is associated with Jimmy Zizmo, husband to Lefty's cousin Sourmelina and his landlord. Lefty's failure to own his home and Zizmo's questionable influence mean that Lefty is eventually out of a job—but not before he has the opportunity to appreciate the technotranscendent heights of Ford factory culture. The melting pot drama, we are told, takes place at the foot of "a building like a gigantic church organ, pipes running into the sky" (92). Fordism does not just tower above Lefty; it infiltrates his very cells. As Eugenides's narrator explains, the adoption of the assembly line production strategy, whereby anyone can do any job and everyone is replaceable, meant that "people stopped being human in 1913." Some of the workers quit, of course, but others took their places, and since then, Cal notes that "the adaptation has been passed down: we've all inherited it to some degree, so that we plug right into joysticks and remotes, to repetitive motions of a hundred kinds" (95). To live in the twenty-first century is already to be some form of transhuman being, a product of cultural determinism that is as inheritable as blond, wavy hair.[25]

One might well ask, "What about black, frizzy hair?" Using a reintroduction strategy akin to Smith's, Eugenides allows Zizmo to exit stage right, apparently plummeting through thin ice while on a hooch errand to Canada, but then brings him back as W. D. Fard, the actual founder of the Nation of Islam who appeared in Detroit in 1930 with a mysterious background.

Alongside Desdemona, we listen as Eugenides uses Fard to tie industrialism to racism. Playing a key predecessor to Elijah Muhammad and Malcolm X, Zizmo uses his sermons to indict an ancient "GOD-SCIENTIST" named "YACUB" who proved too smart for his own good. Yacub tried to "CREATE A RACE OF PEOPLE COMPLETELY DIFFERENT FROM THE ORIGINAL PEOPLE—GENETICALLY DIFFERENT" that he imagined would be able to "DOMINATE THE BLACK NATION THROUGH TRICKNOLOGY" (154). Using a mysterious form of artificial selection that "GENETICALLY CHANGED THE BLACK MAN, ONE GENERATION AT A TIME," Yacub created the white race by making blacks "PALER AND WEAKER, DILUTING [THEIR] RIGHTEOUSNESS AND MORALITY, TURNING [THEM] INTO THE PATHS OF EVIL" (155). Eventually, it would become sadly obvious that the God-scientist's creation shared his abilities: "BY DESTINY, AND BY THEIR OWN GENETIC PROGRAMMING, THE WHITE RACE EXCELLED AT TRICKNOLOGY" (160). This is Zizmo's explanation for why white Americans would continue fostering racism, not only against African Americans but also against American Indians and other minorities.

Zizmo might seem far from a trustworthy voice, but the novel generally verifies W. D. Fard's assertions. There are multiple allusions to the assimilation of American Indians: the Stephanides's pre-*Middlesex* home, for instance, is located "on Seminole" (223), a tribe effectively whitewashed out of existence by colonialism, a reminder that "this whole country's stolen" (241). *Middlesex* spends more space on racism against African Americans, using a young Callie to expose her father's fears of African Americans and at least partly justifying the "guerilla uprising" of 1967 Detroit as "the Second American Revolution" (248). Classism is also evident here, with the novel demonstrating how the point system used to find appropriate buyers for suburban Detroit real estate functioned as a kind of positive eugenics and repeated the immigrant quota system that Cal's grandparents faced on Ellis Island. In short, Eugenides reaches beyond the problem of individual racist and classist behaviors to expose systemic injustices. Beneath the sublime architecture of twentieth-century technoculture, *Middlesex* shows how it was too easy to overlook "workers sleeping on the streets because of the housing shortage, and the ghetto just to the east, a thirty-square-block area bounded by Leland, Macomb, Hastins, and Brush streets, teeming with the city's African Americans, who weren't allowed to live anywhere else." As with Cal's genetic mutation, "the seeds of the city's destruction" were planted long before their effects became easily visible (88).

Of course, the greatest social injustice that *Middlesex* tackles is discrimination on the basis of sex and gender, especially due to nontraditional sexual orientations and complex gender identifications. If immigrants "cannot escape their history any more than you yourself can lose your shadow," neither can intersexed people like Cal escape their biology (385). Yet their genetic inheritance is not everything: it only becomes a gender identity at specific points in time and place, through interaction with attitudes ranging from affirmation to apathy to condemnation. For Eugenides's novel, neither a masculine linearity nor a feminine circularity is sufficient. What is required is a hybrid spiraling forward, so that new generations' differences from their forebears are as crucial as the similarities. This means that in terms of gender and sexual identity, as with race and class identity, *Middlesex* comes down—where else?—in the middle, recognizing the reality of genetic influence but also resisting the notion that genetics determines all. As such, Callie remains a part of Cal, with the narration suggesting that American culture is yet to figure out how to allow them to coexist or for Cal to be genuinely multigendered, rather than simply covering over feminine behaviors with masculine ones.

In sum, when the elderly Desdemona notes near *Middlesex*'s conclusion that the spoon she used to predict Cal's gender before birth was ultimately correct, and Cal responds, "I guess so," much remains unsaid (527). My argument is that the novel suggests not that the spoon was accurate or that it was wrong but that the problem is our binary approach to gender itself, one now exacerbated by a simplistic approach to genetics. What *Middlesex* demonstrates is that rather than more coding of behavior—as with Dr. Luce's testing of Callie to see if his patient "wrote in a linear, masculine way, or in a circular, feminine one" (20)—American culture needs what another intersexual character named Zora says is "what's next" (490). This is genetic realism's equivalent to the unexpected possibility represented by Jodahs in Butler's *Xenogenesis*. Eugenides is seeking not just a polite compromise but a dynamic form of hybridity that instead of gluing together opposites allows them to become new things entirely. Thus the final page's metafictional hint that although the Stephanides "ruined [Middlesex] with our colonial furniture, it was still the beacon it was intended to be"—it was "a place designed for a new type of human being, who would inhabit a new world" (529). This is exactly what Eugenides's novel seeks, an authentically new world, one shaped from the past but open to many unpredictably hybrid ways of being.

I take unusual liberties in speculating about Eugenides's compositional intentions in this chapter partly because of how explicit he has been in published interviews and partly because of what he has told me directly. As an example of his clarity about the active relationship between genes and environment—a vision very much consonant with an epigenetics movement that was yet to become well known—consider this moment from an interview with Jonathan Safran Foer just before *Middlesex*'s publication:

> I grew up in the unisex '70s. The heyday of nurture. Everyone was convinced that personality, and especially gender-specific behavior, was determined by rearing. Sexologists and feminists insisted that each child was a blank slate and that rearing determined gender roles. Now everything is reversed. Biology and genetics are considered the real determinants of behavior. Having lived through the demise of the first oversimplification, I suspect the imminent demise of the current one. Right now we exaggerate the role of genes in controlling our destiny. As Cal says, the ancient Greek notion of fate has today been carried into our very genes. But that's not the way it works. Genes and environment interact during a specific, crucial developmental period. They coauthor the human being. Biologists understand this, but the culture at large still doesn't, quite. (79)

Eugenides explains that an overemphasis on nature has replaced the exaggerations about nurture that shaped his own upbringing. However, *Middlesex* does not simply urge compromise or balance. Eugenides says that "between the alternatives of nurture and nature, I argue for a middle place. That's one of the meanings of the title, obviously." Then he goes on: "But the *Middlesex* I'm talking about is not only a third gender category. It also represents a certain flexibility in the notion of gender itself. It's a very American concept really. It's a belief in individuality, in freedom. I think we are freer than we realize. Less genetically encumbered."

Eugenides's reflection represents the promise of the genetic realism that has continued to emerge in the years since *White Teeth* and *Middlesex* were published. As *Never Let Me Go* and *Orphan Black* also suggest, it is possible to embrace the genome age's capacities to expand self-understanding and new forms of healing without succumbing to the determinism that would reduce the soul to a digital formulation. Still, this requires a constant awareness that genetic biology is not the only force involved, that its significance constantly shifts according to the assumptions of new narratives. Scientific data will always require interpretation, a process often lost in rhetoric that too easily separates facts from meanings. As examined in the remaining two chapters of *Editing the Soul*, some of the most balanced representations of new genetic discoveries and their potential applications are now coming via genetic metafiction, a narrative form into which both genetic fantasy and genetic realism are capable of evolving. We begin in chapter 4 with post-apocalyptic fantasies that allow genetic fiction to say as much about its own capacities as those of new biotechnologies.

4.

Serpent Women, Prophets, and Satire in Genetic Metafiction

It is community theater at its hokey best. Alison Hendrix, *Orphan Black*'s uptight soccer mom, has fallen into the main role in a local adaptation of the grisly musical *Blood Ties*. After the production loses its original lead to a self-strangulation involving a scarf and a kitchen sink trash disposal—an accident Alison witnesses but chooses not to interrupt—the show's most frenetic character gets her big acting break. In one scene, our sappy heroine sets up the next song with this complaint: "Well, I know nobody likes to touch a dead body, but I was really hoping you'd help me clean up the blood." Then the mop-wielding amateur cast launches into an enthusiastic march: "Now we must heed the call / cleaning the brains off the wall / the task may be unpleasant / but it's ours!" This is cringe-worthy enough in rehearsals; when Alison tries to drink her way through opening-night butterflies, her face-plant off the front of the stage becomes full-fledged theatrical disaster. For all its hilarity, though, the scene ends up foreshadowing actual events. After Alison's husband, Donnie, inadvertently shoots geneticist Dr. Leekie in their car, she literally helps "clean up the blood," first in the vehicle, then in their garage. As this chapter demonstrates, the scene's mixture of satire and self-reflexivity points to genetic fiction's gradual shift from relatively straightforward fantasy and realism toward metafiction, a self-aware literary form that comments (sometimes subtly, sometimes explicitly) on its constructed nature.[1]

Another text examined in the last chapter, Jeffrey Eugenides's *Middlesex*, also hints at this self-interpretive drive. The novel's metafictional gestures are first evident in Lefty and Desdemona's romantic performance aboard the ocean-crossing ship by which they escape a burning Smyrna and head for a new life in Detroit. To hide the fact that they are brother and sister, they construct an elaborate ruse in which they seemingly meet as strangers aboard the vessel. The key is that their fiction, which climaxes in a vigorously rocking lifeboat, does not deny reality so much as consciously reconstruct it. As their grandchild Cal narrates it, "They passed the voyage playing out this imaginary flirtation and, little by little, they began to believe it. They fabricated memories, improvised fate . . . it wasn't the other travelers they were trying to fool; it was themselves" (68). This is only the beginning of how *Middlesex* imagines fiction as a tool for rewriting one's circumstances, for changing one's own orientation to reality. That capacity also emerges when Lefty sees Electric Park through his car windows late one night. Setting aside the fact that "the amusement grounds should be closed," the narrator decides to embrace the fantastic "for my own purposes" (110), employing "a moment of cheap symbolism only" (110–11) before again "bow[ing] to the strict rules of realism" (111). Fiction breaks determinism, genetic or otherwise, because as Father Mike notes, "People live by telling stories" (179). The novel's most positive illustration of this survival mechanism comes via Cal's mother and her World War II newsreel experience. Though she has promised to marry the dour Father Mike, her imagined glimpse of Cal's eventual father changes her mind. The decision holds even when she realizes that the man she saw on screen was not actually Milton: "It doesn't matter, however. She has seen what she has seen" (193). The fiction, however brief and insubstantial, gives her the perspective to recognize her feelings about reality, to make a choice rather than merely submitting to others' pressure.

These gestures by *Middlesex* and *Orphan Black* indicate how the shift from genetic fantasy and realism toward metafiction depends on characters and narrators modeling conscious choices to suspend disbelief. In both episodes, sexuality and death rub up against one another, a convergence that grows even more evident in this chapter's primary texts, Brian K. Vaughan and Pia Guerra's long-running graphic novel *Y: The Last Man* and Margaret Atwood's *MaddAddam* novels (*Oryx and Crake*, *The Year of the Flood*, and *MaddAddam*). In setting the fluidity of gender and sexual desire beside the fragility of human life, these postapocalyptic tales blur the lines between spirituality and science and between animals and humans. Resisting the excesses of absolutism indulged by both militant scientism and the religious fundamentalism to which it often overreacts, *Y: The Last Man* and *MaddAddam* represent the cutting edge of genetic fantasy's evolution into metafiction. By imagining new species and epistemologies, they demonstrate that genetic metafiction is about not just weighing the potential of emerging biotechnologies but directly assessing its own capacity to shape cultural responses. Eventually, by focusing on the precise operation of Atwood's biotechnological novums, we will see how much this emerging subgenre can reveal about our approaches to gender, religion, and narrative.

Postapocalyptic Pseudoscience

In the opening pages of *Y: The Last Man*, nearly every male on the planet is struck down by a virulent plague that induces nearly instantaneous death. As in Mary Shelley's lesser-known second novel, *The Last Man* (1826), a mysterious outbreak decimates life on Earth, but instead of destroying both sexes, Vaughan and Guerra's version only wipes out organisms with a Y chromosome, except the protagonist and his monkey. Unlike Shelley's apocalypse, there is a regular comedic inflection to this tragedy in its main narrative and especially in its *mise-en-abîmes*. The most prominent of these is a play written for the women of a small Nebraskan town, whose leader declines *The Pirates of Penzance* and *Glengarry Glen Ross*, requesting instead a theatrical adaptation of *As the World Turns*. Presenting the traveling troupe with six months of *Soap Opera Digest* and asserting that the only musical she knows is *Moulin Rouge!*, which she "didn't really care for," this Stepford wife is mostly concerned that the production not include swear words. This is difficult for the thespians to accept; their work is already limited in its production values, with stage lights depending on generators and leftover gasoline and with women having to glue on fake beards to play the male roles. Moments before the conversation, in fact, two performers had ended a *Penzance* rehearsal by arguing over a fight scene, with one calling its violence "gratuitous" and suggesting that their new all-female audiences are "more interested in drama and romance than punching and kicking," while another objects that this is precisely the sort of "myth" about gender that needs "exploding" (2:128). The meeting with the town's narrow-minded representative, though, overturns any assumption that sexism comes only from men. Turning away with a quiet aside, one actress chuckles, "Exploding myths about gender, eh?" Her peer smirks and says, "Yeah . . . KA-BOOM" (2:131).[2]

Such combinations of humor and urgency about social justice characterize the independent comic's entire six-year run. Its protagonist, Yorick, reincarnates the deceased jester who provided the famous skull in *Hamlet*, the same character referenced in Cosima's brain-wielding monologue in *Orphan Black*. Like his Shakespearean forebear, presumably, Yorick simultaneously faces and represses tragedy via his comedic persona's blithe obliviousness. A couple of issues into the plot, after announcing his survival to his mother—now a White House official—he goes in search of a Boston-based bioengineer who might help discover the plague's causes, Dr. Allison Mann. In addition to Ampersand, the male monkey he adopted just before the plague struck, Yorick's traveling companion is "355," an African American NSA/Secret Service type assigned to protect him. She insists her real name is classified; he responds, "And if you'll be my bodyguard, you can call me Al?" (1:89). Undiscouraged when she misses the Paul Simon reference, he continues prattling about how close he and his sister, Hero, are: "Luke and Leia . . . um, minus the French kissing." Then, looking beyond the panel's edge, he is taken aback: "What the hell is that?" (1:91). The answering full-page spread emblematizes the graphic novel's tragicomic tone: the most phallic of national monuments has been transformed into a memorial for mankind (fig. 15).

Such wistful moments appear throughout *Y: The Last Man*, and their purposes reach well beyond entertainment. At first, we might simply enjoy the protagonist's pluck, but it isn't long before his wisecracks start to betray a jocular denial of reality, a too-easy willingness to ignore what he does not understand. When Yorick hears Dr. Mann confess that a new character reminds her of an ex, for instance, his reaction screams heteronormativity: "Hold on, *she* reminds you of an ex-*boyfriend*?" He has been traveling with the geneticist for a year, but it requires a pause across multiple panels for him to grasp Allison's meaning. As she regretfully observes, "I suppose we can add *gaydar* to the extraordinary number of common senses you seem to lack" (2:185). Nor does that ability emerge quickly: Yorick is still shocked, even embittered, upon discovering Mann and 355 in bed much later in the narrative.[3] Decades on, however, we are allowed to see that some of the lessons about gender and sexual preference have finally set in. The graphic novel's concluding pages reveal that the elderly Yorick might not be thrilled about cloning as a way of bringing back the species, but he quickly cuts off his young doppelgänger's joke that "we spend nine months trying to get out of a woman and the rest of our lives trying to get back in." He responds, "Do yourself a spectacular favor and stow that frat-boy horsecrap with a quickness. Girls aren't a game. Not one that you can win" (5:274). The reader sees that this is even more true than Yorick knows: the woman he had asked to marry him just before the plague hit, and for whom he spends most of the story searching, ends up finding happiness in the arms of his sister—a character the graphic novel cheekily names Hero.

Fig. 15 The tragicomic reveal of the memorial for mankind in *Y: The Last Man*, deluxe ed, 1:92.

Vaughan and Guerra's insights about gender and sexuality go much further, though, than urging respect for women and challenging homophobia. Part of the graphic novel's power is in illuminating pitfalls on both sides of the political spectrum, criticizing a hateful mutation of second-wave feminism just as incisively as Yorick's thoughtless overgeneralizations. The story's postcollapse "Daughters of the Amazon" are well beyond bra burning; they require initiates to undergo a single mastectomy to demonstrate their loyalty. Learning of Yorick's existence, their leader hunts him down, not in hopes of renewing the species, but in order to eliminate men once and for all. To the women protecting him, Victoria demands in the most anthropocentric terms possible, "Are you animals or are you *women*?" (1:206). She reminds these escaped convicts that they were once assigned terms twice the length given to men for the same crime, that while they were imprisoned for stealing to feed their families, men who embezzled billions of dollars went free, and that "justice is a woman with a *sword*" (1:207). Yorick decides that the only alternative is to surrender himself, allowing his militant feminist tormentor to quote Shakespeare—"Alas, poor Yorick"—and tell her "sisters" to observe "the fall of man" (1:217). Before she can execute him, though, one of the former convicts fights back. Victoria collapses with an ax embedded in her forehead, and Yorick's journey continues.

This confrontation with a violently reactionary feminism earns *Y: The Last Man* its even more expansive rebukes of male sexism. In Jordan, the story points out, one quarter of murder victims are "women killed by male relatives who simply *accuse* them of adultery or . . . or 'fornication'" (1:22). Some of the most poignant panels require no words at all: the plague's spread leaves the Tokyo Stock Exchange without a person standing; the pope collapses at the Vatican; prostitutes in Amsterdam stare down at suddenly deceased customers; a single shocked scientist remains alive in Mission Control at the Johnson Space Center. The most didactic page in the graphic novel welcomes readers to the "UNMANNED World" with a sobering death toll: "495 of the Fortune 500 CEOs," "99% of the world's landowners," "95% of all commercial pilots, truck drivers, and ship captains," "99% of all mechanics, electricians, and construction workers," "85% of all government representatives," and "100% of Catholic priests, Muslim imams, and Orthodox Jewish rabbis" (1:39). *Y: The Last Man*'s target is not just isolated acts of prejudice or exploitation, then, but larger cultural systems that limit women's possibilities. Certainly there are individual exceptions, including the tellingly named Dr. Mann, Yorick's mother, and his would-be fiancée, Beth (an anthropologist earning her stripes in the Australian outback), but the story testifies that true equality of opportunity remains elusive.

How does *Y: The Last Man* explain ongoing disparities in wages, respect, and opportunities? Vaughan and Guerra's graphic novel implies that beneath many patterns of contemporary sexism lie foundational misunderstandings of religion and of genetics. From its opening issues, the comic indicates its concerns with religious and mythological influences on gender. Note, for instance, the witty pattern of curses that reference Yorick's pseudomessianic role. "Oh my God," Dr. Mann reacts upon first meeting our protagonist, and Yorick cracks, "I get that a lot" (1:123)—even if she's actually looking at his monkey, Ampersand. When one of the convict women finds Yorick unconscious after being thrown from a train, a full-page spread spotlights her reaction: "Thank you, Jesus" (1:154). These could seem routine exclamations, but as they accumulate, it becomes hard not to chuckle at the double entendres. Yorick has no idea that an Israeli soldier's crosshairs rest on him as he reacts to the prospect of more medical tests with "Jesus, kill me now" (2:58); when male astronauts descend from the International Space Station, Dr. Mann comments, "It's raining men," and 355 responds, "Hallelujah" (2:87); on telling Yorick that it was antibodies in his monkey's feces that saved him, Dr. Mann calls it "fitting that the salvation of the Y chromosome" should come in this form, and he replies, "Salvation? Jesus Christ, he's a monkey, not . . . not Jesus Christ!" (3:149). There are many such instances, and their collective impact is to highlight how the protagonist's mysterious biological exceptionality makes him as much a savior figure to some women as he is a threat to others. These are not just superficial jokes; they call attention to Yorick's and other characters' hunger for a deeper sense of purpose. As he puts it to the person he trusts more than any other, 355, "Can you believe I honestly used to think there was a reason I was still here? Divine intervention, fate, fucking magic. . . . There had to be some larger-than-life explanation . . . but now I know it was all just a crap shoot. Motherfucking literally" (3:163). For a time, this seems to be his—and the graphic novel's—answer to the mystery of his survival: not some kind of transcendent role he has been called on to play but the same sense of cosmic randomness that terrorizes Samad in *White Teeth* and Lefty in *Middlesex*.

Ultimately, though, *Y: The Last Man* offers something rather different— a mixture of cosmic spirituality and pseudoscience that crystallizes only in its conclusion. The graphic novel hardly propagates traditional organized religion; like many post–World War II fictions, it criticizes Catholicism especially directly. However, it does so from a much more informed, insider perspective than is often the case. For instance, Yorick meets another Beth, quite distinct from the one in Australia he intends to marry, at an abandoned church near San Francisco. She turns out to have been a theology major at Georgetown, where she wrote a thesis on Agnes Snoth, who, as the text explains, was burned alive in the 1500s along with three other women for preaching against auricular confessions, arguing that "it was sinful to ask a man for what only God can grant" (3:16). Once this other-Beth has tried to kill the intruder and he's revealed himself as male, their relationship swiftly turns intimate. Not long after their tryst in the church graveyard—which the comic figures as Eden redux (fig. 16)—they are interrupted by more Daughters of the Amazon, one of whom intends to burn

Fig. 16 Individual issue cover art emphasizes the retelling of Genesis in *Y: The Last Man*, deluxe ed., 3:29.

down the church as a "disgusting shrine to male hegemony" (3:38). Apparently unworried about accusations of excessively convenient plot developments, Vaughan and Guerra make the attacker a theology major too, this time from Berkeley, and demonstrates that ideology is not just a matter of education. The Amazon woman cites the church's denial of women's reproductive rights and its complicity with nuns' rapes at the hands of their own priests, but other-Beth responds that perhaps the nuns had it coming, considering the abuses of even relatively recent projects like the Magdalene asylums. As she puts it, "The Church wasn't fucked-up because it was run by men, it was fucked-up because it was run by humans" (3:41). Stealing a page from Octavia Butler, other-Beth redirects rage at the opposite sex by targeting a broader problem with abuse of authority.

What makes the scene work is that other-Beth's theological argument does not win the day so much as her fighting skills and Yorick's training as a magician. Projecting his voice and hooding himself beneath clergy robes, he plays a levitating Yahweh who successfully guesses one intruder's history of sexual abuse and quotes *Pulp Fiction* at another: "You will *know* I am the Lord when I lay my vengeance upon you!" (3:44). The Amazon woman recognizes the ruse, but it still provides sufficient distraction for other-Beth to disarm her. In other words, real lives are saved because Yorick can quote a fake Bible verse and simulate a miracle. And the same kind of fantastic power is invoked in Yorick's other intimate encounter during his western pilgrimage, his suicide intervention at the hands of one of 355's former colleagues. Employing an aversion therapy she credits to Benjamin Franklin and the Marquis de Sade, "711" explains later that "it's based on the idea that your sexuality and mortality are indissoluble elements" (2:230). In the midst of the experience, Yorick knows only that he is handcuffed to a bed and that an attractive dominatrix is forcing Viagra down his throat, mounting him, and demanding that he perform, but that when he starts to submit, he nearly gets drowned. 711's goal is neither to have sex with Yorick nor to murder him but to help him rediscover his desire to live. Once that occurs and he throws himself out of the water, the ploy is over—she returns to her former identity as a demure hermit and he is left to contemplate the death drive that has dogged him since the plague began. As in the church scene, the alternate reality provided by a theatrical ruse breathes new life into Yorick, leading him to ponder that neither sex nor death are the ultimate ends they sometimes advertise.

It is not just Yorick whom the graphic novel suggests needs aversion therapy, however: so do its *readers*. We too are confronted with a sometimes horrifying fiction featuring rampant sexual objectification and the potential collapse of humanity and then asked if we care. *Y: The Last Man* becomes a vivid demonstration that *Homo sapiens* exhibit the same death drive at a species level that we see in the story's individual characters. Guilt motivates not just Yorick but also Allison Mann, who believes she has caused the plague by attempting to clone herself and wants only to stop it so she can "kill [her]self in good conscience" (1:121). Her father is unapologetic about his own experiments and is fully convinced that male humanity must disappear, nearly killing Yorick before Allison intervenes. Then there is the graphic novel's most dangerous antagonist (beyond the leader of the Amazons), an Israeli commando who expresses disappointment upon her introduction in the first issue that she has never been fired on during street skirmishes in the West Bank. By the story's conclusion, as Yorick confronts Alter in Paris, he finally recognizes her own death drive—remarkably, given that she has just killed the woman he loved. Held at gunpoint, Alter eggs him on, telling Yorick she also killed his mother, but then it hits him: "That's what this has been about. The whole time. You've been trying to commit suicide, too" (5:251). Given the chance for revenge, he knocks her unconscious, tosses his gun before her troops, and echoes words once spoken on Mount Calvary, "It is finished." His declaration is merely "Enough" (5:255), but if the symbolism around his simultaneous surrender and act of mercy is unclear, the next page provides one more exclamatory curse from the original Beth upon finding him at the Fontaines de la Concorde: "Jesus Christ" (5:256).

What *Y: The Last Man* suggests, like Butler's *Xenogenesis* trilogy, is that sexism and hierarchical ideologies, especially religious ones, are often mutually constitutive. Moreover, when offered no appropriate outlet, sexual desire can turn murderous or suicidal. We have seen in earlier chapters how often genetic fiction emphasizes love's dependence on openness to an other, including all their unpredictable, discomfiting differences, and in Vaughan and Guerra's tale, the main characters also discover the necessity of this radical vulnerability, especially Allison Mann. She once tried to serve as birth mother to her own clone in order to get full credit for the scientific innovation, attempting to avoid the fate of double helix codiscoverer Rosalind Franklin, whom she views as a mere "footnote" in the historical record. She emphasizes that this let her "eliminate the most painful part of the process," and when her student asks, "What's that? *Sex?*," she responds, "Love" (4:256). The further we reach into her past, the more it brings to mind Rachel Duncan of *Orphan Black*: Allison Mann also had a scientist father who found her disappointing, so much so that he resolved to use cloning as a second attempt at parenting "her." Ultimately this is the experiment Vaughan and Guerra blame for the plague: when the first clone of Dr. Matsumori's daughter is born, a "shock wave" travels the globe at the speed of light, 186,000 miles per second. The premise of *Y: The Last Man*, the novum on which the narratives about parental abandonment and the discovery of authentic love depend, is a nearly mystical link between genetics and global ecology.

However implausible, the specificity of this explanation is fascinating. *Y: The Last Man* draws more heavily on fantasy than realism, but in blurring that line, it moves toward genetic metafiction. Like *Gattaca*, the graphic novel is an often profound, aesthetically impressive critique of eugenic thinking, yet its premise requires a massively exaggerated genetic determinism. In *Gattaca*'s storyworld, the problem is a nearly complete denial of environmental factors or individual agency in shaping genetic expression. The comic's future, by contrast, features a radical overestimation of environmental factors. Vaughan and Guerra's project incorporates an ecological vision much like that of James Cameron's *Avatar* (2009), one that works more as an allegory about social justice than as an attempt to unite scientific and humanistic values. Its "morphogenetic" explanation goes well beyond understanding genetics and environments as deeply interwoven factors to construct a literal Gaia-like force that consciously shapes life on Earth.

In a flashback, we learn that even when Dr. Mann was merely a precocious girl and heard the theory of "morphic resonance, the socio-biological interconnectedness of species," she rejected it. Told that "our genes are receivers capable of transmitting and obtaining information through the unseen 'frequency' that unites all life on this planet, like the invisible bond that holds together atoms of a molecule" and therefore that "homework is easier to do in the evening because your fellow pupils will have already raised the collective consciousness we've come to think of as 'instinct,'" she immediately called the idea "stupid" (5:32). Her father acknowledged that she was probably right, but decades later, he relies on this pseudoscience to explain why his human cloning project killed nearly all the males on the planet. Since his explanation is never contested, *Y: The Last Man* seems to support his view. "All life in the universe may not be connected," Matsumori says, "but all life on our planet is." Then he goes much further, attempting to render a figurative reality as if it were a measurable, literal one. Claiming that "we are surrounded by the biological equivalent of electromagnetic fields," he describes groups of chimpanzees learning from each other despite being separated by thousands of miles, and says the same thing happened when he "unlock[ed] the secret of asexual human reproduction." As he understands it, "The Y chromosome has been rationally self-destructing for hundreds of millions of years. It used to contain thousands of working genes, but was whittled down to just a few dozen even before the plague. Men have long been a necessary evil for the continuation of this species, but the moment that evil became obsolete, nature righted its course. I was merely the trigger that set off a time bomb that's been ticking for millennia" (5:68–69).[4]

On one hand, then, Vaughan and Guerra's graphic narrative is a brilliant step in the evolution of genetic fantasy. Scientific details aside, it insightfully demonstrates how consumers of genetic information can be duped, as when Dr. Matsumori's henchwoman, Toyota, tells 355 in the midst of their fight scene that he's "gonna help me live forever" (5:70). When 355 tersely replies, "You're an idiot," she acknowledges, "That's what I told Dr. M, but then he showed me all the test-tube brats he engineered. Says he's only gonna make copies of certain girls, and I'm one of them. A thousand years from now, there will still be a whole line of women exactly like me walking the planet. I'll never really die" (5:70–71). This is the old carbon-copy myth, alive and well, yet the graphic novel exposes its fraud, not least when 355 kills Toyota at the scene's conclusion. Even if Dr. Matsumori had fulfilled his promise, she would never have experienced it; the being she understands as her self, the person we know as Toyota, very clearly *does* die and would have in any case. The graphic novel powerfully exposes this limitation on many posthumanist pitches; at the same time, *Y: The Last Man* writes nature above humanity as a kind of mystical, unbridled force barely distinguishable from Paley's intelligent designer. If nature makes males expendable, we are to believe it's the end of men, virtually instantaneously.[5]

At the same time, Vaughan and Guerra's tale reaches beyond genetic fantasy to metafiction.[6] We have noted the dual humor and poignancy of Yorick's messiah complex, his and other characters' interwoven sex and death drives, and the use of *mise-en-abîme* to highlight tensions within the central plot, as with the play performed by the traveling actors. I can now add that it is not just that *Y: The Last Man* holds up a more honest mirror to those who want art to affirm their insularism and superficiality; it uses this drama to comment on its own role in literary and cultural history. The comic's self-reflexivity does not require taking itself overseriously: as one of the actresses complains during a rehearsal, "If there's one thing I hate, it's crappy works of fiction that try to sound important by stealing names from the Bard" (2:149). This aside to readers (in a graphic novel whose main character is named Yorick) simultaneously lightens the mood and raises our expectations: the narrative is defying those who would dismiss its medium as pulpish, adolescent escapism. Instead, Vaughan and Guerra use the story within the story to confront their protagonist with the possibility that the best option for any surviving last man on Earth would be "committing suicide, and letting the women save themselves" (2:166). Yorick rejects the assumption, but it haunts him throughout the story. It is on his mind as he undergoes 711's suicide intervention, and it is what she alludes to with the parting gift of her copy of *War and Peace*, urging him not to "skip ahead" because "endings have to be earned" (2:233).[7] Put another way, there's no telling how he might best contribute to humanity's salvation; there are too many contingencies, too many choices left to be made. Rather than pretending his tale is fated to become another "crappy work of fiction" and collapsing under the weight of survivor's guilt, he must play his role through to the end.

Featuring myriad literary and pop culture references, the comic never stops reflecting on its own potential.[8] The metafictional components become most salient when we catch up with the traveling thespians just before the story's climax. In an issue entitled "Tragicomic," we learn that they have made it to Hollywood and are busy "*appropriating* the trappings of male-dominated cinema, and *subverting* them to make the first truly female action hero" (5:128). Or at least, that's the goal: while the play they wrote as a rejoinder to the latter-day Stepford community eventually became a nationwide hit, film proves a tougher medium. The Berkeley theology major disarmed by other-Beth in the church scene has left the Daughters of the Amazon and is now playing one in the film, but she walks off the set, condemning it as merely "another garbage action movie, exactly what the *patriarchy* used to churn out" (5:127). The company's heroine then follows suit, saying she's going back to Broadway because she'd "rather do something great for a few people than dumb everything down just to be palatable to the masses" (5:129). These desertions kill the project, and the main writer and director are soon back on the road, lamenting the absence of Oscar nominations for female directors even before the plague. When they are carjacked by a gang of preadolescents calling themselves "the Last Girls," they are left with nothing but their writing and art skills, but the holdup proves a blessing in disguise. One of their assailants drops a comic entitled *Heartstrings* that reeks of sentimentality, but it inspires their foray into a format with "all the advantages of film and none of the drawbacks"—"the cheapest way to get [their] unfiltered vision into as many hands as possible!" (5:136). Vaughan and Guerra's own goals in *Y: The Last Man* could not be stated much more clearly: "We could create something new, something that challenges our audience at the same time it's helping them *escape*. Artists are supposed to hold a mirror up to society, but ours could be a . . . a fucked-up *funhouse* mirror!" (5:137).

On the next page, readers sample the comic the fictional playwrights produce. Entitled *I Am Woman* (rather than *I Am Legend*), it features a lone woman and her trusty mare traversing a postapocalyptic desert landscape. Two months after all the other *women* died, the heroine's thoughts echo the early pages of *Y: The Last Man* but with the genders reversed: "Who knew the world would crumble so quickly just because 98% of the secretaries and kindergarten teachers died? Who would've guessed that society would collapse without nurses or maids or waitresses or freakin' *librarians*?" However, the narrative is quick to recognize that women's importance far exceeded these roles, not to mention those of "mothers and wives and sisters." In fact, "they were the only thing preventing the boys from beating each other into oblivion and then raping the corpses" (5:139). Eventually a copy makes its way to Yorick, and while riding the Trans-Siberian Railway toward Paris, 355 sarcastically inquires about the "literary masterpiece" he is perusing. He says the comic has become "all the rage back in the States" (5:141) and explains that it's not a "capes-and-tights book" but a "quasi-feminist sci-fi thing. Very po-mo" (5:142). Of course, the same can be said of *Y: The Last Man*: the sequence's cumulative effect is that the genders need each other. The situation could be reversed, Yorick could be the last woman alive, and the significance would be very similar. He is himself not just because of his genes, not just because of his gender, and not just because of his culture but because of their complex interactions. His identity is the result of the stories he has lived through, both as a reader and as a character, and this is why one of his last exchanges with 355 has him jotting a list of novels he hopes she might pick up one day, including Smith's *White Teeth* and Atwood's *The Handmaid's Tale* (1985).

Vaughan and Guerra's graphic novel troubles easy distinctions between the genders, oversimplifications about genetic determinism, and underestimations of storytelling's power. It also joins many of the best genetic fictions in a complex assessment of new biotechnologies: their expansion is a given in both the pre- and the postapocalyptic worlds of the graphic novel, so the challenge is to avoid inflating *or* discounting their significance. *Y: The Last Man* refuses to defend genetic determinism like that championed by Allison Mann's father, and it allows the elderly Yorick several parting shots about how "new strains" (5:281) of his clones sound like diseases rather than people. At the same time, it declines to make science and technology into unnecessary enemies. Human clones seem a necessary part of postpandemic culture, and the graphic novel's final, wordless panels inscribe them within the beauties of ordinary life, not just the sterile spaces of the research laboratory. Even where these doppelgängers live and work together, they differ; one of Mann's clones stands out particularly for her punk look. As Yorick summarizes, "You can photocopy Mann's brain as many times as you want, but without her asshole dad pushing those girls every step of the way, you'll never have her mind" (5:287). This is precisely the epigenetic awareness toward which we saw *Orphan Black*, *Middlesex*, and *White Teeth* lean in chapter 3, and it is even more central to the genetic fantasy-turned-metafiction on which this chapter focuses most extensively, Margaret Atwood's *MaddAddam* trilogy.

Since the earliest genetic fantasies, myths of carbon-copy human cloning have supposed that what the factory enabled for industry, the lab might create for genetic engineering. This is partially true: our species has grown fixated on the power of exact replication, and as *Middlesex* dramatizes with Ford's factories, the assembly line's appeal is predictable, controllable results without inconvenient influences like subjectivity or emotion. The assumption has been that genetic manipulation would operate with equal efficiency, but there are often vast distinctions among genes as cultural metaphors and as scientific referents. Genetic fiction is not just a means of illustrating biotechnological achievements but a retrospective lens that heavily influences the culture into which future tools emerge. These stories have regularly encouraged shifting from the exaggerations of unbridled genetic determinism toward the more complex vistas of epigenetics and other, richer syntheses of biology and environment. Sometimes such tales have been relatively allegorical and fantastic, other times more directly realistic, but many are now becoming increasingly self-conscious. Look back once more to *Middlesex* and an evocative allusion Cal makes to the seventeenth-century Preformationists. As he explains, they "believed that all of humankind had existed in miniature since Creation, in either the semen of Adam or the ovary of Eve, each person tucked inside the next like a Russian nesting doll" (199). Eugenides is effectively describing the carbon-copy clone myth, and his novel can be read as anticipating the epigenetics movement's role in debunking it. Instead of the Preformationists' ontogeny-recapitulates-phylogeny vision—in which early scientists expected dissection of a silkworm to "reveal what appeared to be a tiny model of the future moth inside" (199)—*Middlesex* suggests that we find ourselves only through an unpredictable, dynamic interaction of inheritance and experience.

As it turned out, Eugenides's nesting doll metaphor foreshadowed a key image in Margaret Atwood's most complex work of speculative fiction, the *MaddAddam* books. Granting the obvious virtues of *The Handmaid's Tale*, and with respect to other wonderfully ambiguous utopian-dystopian works like Ursula K. Le Guin's *The Dispossessed* (1974), this trilogy flits back and forth more nimbly than any preceding fiction between humanity's near-term biocultural circumstances and a postapocalyptic future. If the very hinge of Eugenides's novel, its eleventh chapter, imagines how "everything comes out of an egg," Atwood literalizes the notion of predetermined consciousness through her Crakers, the transhuman beings who emerge from a hermetically sealed dome they call "the Egg." They are designed by a young scientist (self-named Crake) to repopulate the Earth after a plague that proves even more devastating than the unintentional pandemic imagined in *Y: The Last Man*. Crake's genocidal contagion has eliminated most traditionally reproduced human beings; he likely inoculated many of those who remain, with or without their knowledge. Several were part of the unconventional God's Gardeners cult focused on in the trilogy's second book, and they understand the plague as a "Waterless Flood," a recapitulation of the Genesis narrative that aims not for a perfect world but for a fresh start. It is a final solution beyond any previously imagined eugenic project, and if the Crakers become a wandering tribe of Israel, these former God's Gardeners and the surviving "MaddAddam" scientists are the postplague world's family of Noah.

On its face, the *MaddAddam* trilogy offers a deceptively simple plot: embittered by unbridled corporate greed, a youthful Moreau figure nearly ends humanity with a bioweapon, creating simple-minded replacements whose genitalia turn blue like vervet monkeys and who will mate with anyone who smells right. That might sound like a screed against the hubris of scientists, the sentimentality of religious types, or both, and there's no denying that the trilogy's most egotistical geneticist, Crake, and its most powerful religious leader, Adam One, share a seemingly innate self-righteousness. Instead of thoroughly idealizing or demonizing various ideological extremes, though, Atwood is up to something far more nuanced. By focusing on a single object that lies at the core of the narrative—the serpent-woman saltshaker hiding the chess bishop and its deadly pharmaceutical contents at Scales and Tails, an upscale gentlemen's club—this section demonstrates how closely related are the tasks of interpreting the trilogy's treatments of sexuality, religion, and biotechnology. Fully grasping what the *MaddAddam* novels are saying about the exploitation of women, the nature of the soul, and genetic engineering requires considering these topics together. Before exploring the trilogy's metafictional insights, we must understand its feminist approach to the sacred aura around new capacities for controlling human reproduction.

In the trilogy's third novel, readers step back in time to learn how the deadly bioforms that became the basis for Crake's plague first escaped the biotech compound where they were manufactured. A high-ranking scientist named Pilar, soon to join the God's Gardeners, lifted six deadly pills from the lab, encasing them within a hollow chess bishop whose head is conveniently detachable.

She gave them to Zeb, younger brother of future God's Gardener leader Adam One, who was fleeing the Rev, one of the trilogy's most despicable characters. This backstory suggests why Adam and Zeb fear the charlatan who raised them and why the older boy eventually attempts to rewrite his father's pious path, while the younger replicates the man's secret violence. A key moment for both characters arrives when their former parent descends on Scales and Tails for a night of discreet upper-class entertainment. With Adam's help, Zeb has traded his disguise as a low-level techie at HelthWyzer West, where he received the deadly game piece, for an equally fake identity as a bouncer at the ritzy strip club. There are layers upon layers here: he has hidden the white bishop and its pathogenic cargo inside a functioning saltshaker shaped like a naked, green, scale-covered woman. It is a three-piece nesting doll of the sort imagined by Cal in *Middlesex*, a mirror for both the contortionists who work at the establishment (who are full of their own surprises) and, more abstractly, for the larger narrative's imbrication of gender, religion, and biology.[9]

To the casual eye, there is no sign that the serpent woman is pregnant, but the "green lady with the bishop up her snatch" (303) is a synecdoche for Atwood's entire project. Encapsulating biotechnological horrors (the pills) behind a religious facade (the bishop) and burying both in an unapologetically objectified and sexualized female figurine, the *MaddAddam* trilogy encourages readers to interpret big biotech with attention to its cultural packaging. Unlike in many biopocalypses, from *The Island of Doctor Moreau* to *Y: The Last Man*, Atwood's mad scientist and his rationalized experiments do not appear on the scene ex nihilo, nor are they vanquished by a triumphant hero. Instead, her parable reaches deep into these characters' backgrounds, examining scientific capabilities and limits via more detailed attention to personal beliefs and sexual desire. This requires constructing continuities of scale between the impossibly small, the familiarly human, and the unthinkably large: like Timothy Morton's concept of the hyperobject, a twenty-first-century version of the sublime, Atwood's vision links phenomena as large as the planetary forces creating climate change and as miniscule as the microbiological spaces of genetics. Both scales call for "serpent wisdom," a God's Gardeners idea so core to Atwood's vision that she planned it to be the title of the second novel.[10]

One layer at a time. First, why are the pills and the bishop buried within a serpent-woman saltshaker? Because Atwood's brand of second-turned-third-wave feminism is committed to detailing how the holy costumes and pageantry of ritual often unduly foster the simultaneous deification and demonization of women. That's what the serpent woman represents in this novel, much like Lilith in Octavia Butler's *Xenogenesis* trilogy or other-Beth hissing at Yorick after seducing him in *Y: The Last Man*: a darker, less innocent, culturally overdetermined Eve. She takes a range of forms in Atwood's novels: here, Crake's plague is delivered around the world by an abused angel named Oryx, whose appeal for both Crake and Jimmy is related to her sexually enslaved childhood. Oryx demonstrates how the myth of the pseudodivine woman must always be greater than the reality and how this blind idealization often enables the worst applications of both religion and biotechnology. Similarly, readers learn near the trilogy's conclusion that Eve One, the first female leader of the God's Gardeners, was Katrina Wu, the entrepreneur who created the Scales and Tails strip club. Appropriately, in a still earlier life, she was the magician's assistant Zeb knew as Miss Direction. In these earlier contexts, she used sex appeal to distract audiences, whereas eventually she hid the higher purposes of Gardener women beneath modest behavior and shapeless dyed bags masquerading as dresses. This is to say nothing of the deceptions of Toby and Ren, middle-aged and teenaged reluctant Gardeners who began by rolling their eyes when told to prepare for the "Waterless Flood" but end up among its few survivors. Unlike many who die at the hands of Crake's plague, they escape the nostalgia that yearns for the world left behind. Instead of looking back and being turned into pillars of salt (or pink goo, in the case of the Rev), they keep stumbling forward. All these women exhibit serpent wisdom: they sense that hyperidealizing the female body is a step toward the objectification of all bodies, and via pretended submission, defiant resistance, or both, they resist any dualism that would separate the soul from its physical incarnation.

If these heroines in the *MaddAddam* trilogy refuse to essentialize gender, other characters embody the trilogy's equally nuanced approach to religion. Certainly there are moments in which Atwood's tongue is stuck firmly in cheek—she must have laughed out loud while writing parts of Adam One's sermons and hymns—but her fondness for her imagined cult also comes through. It is crucial that the bioweapon pills are not hidden directly in the serpentine saltshaker but only make it there via the wily bishop. The chess-piece clergyman represents the trilogy's rejection of much traditional religion and its simultaneous appeal to apophatic theology, a fruit of monotheism that receives too little airtime. This branch of theology relies on negation—what cannot be said about God—rather than the doctrinal confessions of cataphatic theology, which range from the poetry of the Apostles' Creed to the idiocy of Westboro Baptist protest signs. Naturally, most monotheism featured in popular media is cataphatic: those who hate gays list the sins they think God abhors, rarely discussing what they *don't* know about God. By contrast, the God's Gardeners are usually too busy surviving to force their tradition on others, and they are also far more comfortable with the unknown. Consider, for instance, Adam One's claim that "God is pure Spirit; so how can anyone reason that the failure to measure the Immeasurable proves its non-existence? God is indeed the No Thing, the No-thingness, that through which and by which all material things exist; for if there were not such a No-thingness, existence would be so crammed full of materiality that no one thing could be distinguished from another" (*The Year of the Flood* 52). This is not atheism, no more so than a physicist's embrace of dark matter is antiscientific, but a heterodox agnosticism combined with a consistent morality. As Adam One tells Toby when she hesitates to accept a leadership role, fearing that her lack of belief would make her hypocritical, the key is in her actions, not her words.

For some, the notion that such a famous feminist and critic of theocracy as Atwood could be aligning her voice with any form of religion might seem, well, heretical. After all, Atwood became one of the most banned authors in the world after publishing a satire about a future America that abandons both women's rights and any meaningful division between church and state. That represents only a selective reading of *The Handmaid's Tale*, though: Atwood's writing has been deeply critical of religious fundamentalism, with its false pieties and hypocritical slander, but even in that famously contentious novel, there remains room for other varieties of religion. Note, for instance, Offred's sympathetic rendering of the Lord's Prayer, which I would suggest represents Atwood's own reaction to the culture wars, particularly in its conclusion, "I don't believe for an instant that what's going on out there is what You meant" (194). As is also evident in the *MaddAddam* novels, Atwood's feminism is fueled by agnostic humility and apophatic theology alike. Even her most disgusted critic of religion, Crake, ends up amplifying Atwood's questions about God. Though he tries to delete all religious instincts from the Crakers' genomes, Ren's reflection about Crake suggests that he is closely aligned with his creator's perspective:

> [Crake] used to say the reason you can't really imagine yourself being dead was that as soon as you say, "I'll be dead," you've said the word *I*, and so you're still alive inside the sentence. And that's how people got the idea of the immortality of the soul—it was a consequence of grammar. And so was God, because as soon as there's a past tense, there has to be a past before the past, and you keep going back in time until you get to

> *I don't know*, and that's what God is. It's what you don't know—the dark, the hidden, the underside of the visible, and all because we have grammar, and grammar would be impossible without the FoxP2 gene; so God is a brain mutation, and that gene is the same one birds need for singing. So music is built in, [Crake] said: it's knitted into us. It would be very hard to amputate it because it's an essential part of us, like water. (*The Year of the Flood* 316)

Music, language, genetics, theology—through Ren and Crake, Atwood suggests that we take them together or not at all. The material illuminates the path to the immaterial, a route that depends more heavily on good questions than on certain answers. Science and religion are neither necessarily opposed nor reducible to one another; they ask distinct but overlapping questions. Atwood's trilogy implies that when faith and reason are understood aright and when fiction and nonfiction are both capable of their own varieties of truth, very different methods of approaching reality can provide surprisingly congruent insights. From a postsecular angle, the key is to recognize the massive diversity among and within world religious traditions and then to place specific manifestations in active dialogue with reason and refuse to let either category assimilate the other.

A fuller elucidation of that possibility awaits in this chapter's final section, but for now, consider the third layer of the serpentine saltshaker, the pills hidden inside the bishop. If the trilogy's approach to biotechnology is encased within its feminism and its apophatic theology, what are the *MaddAddam* books suggesting about genetic research and applications? After completing the first volume, *Oryx and Crake*, readers often view the story as another cautionary dystopia, one exposing the technotranscendent hubris of its young geneticist entrepreneur and warning just how easily a lone bioengineer might devastate the planet, given enough anger or callousness. In many ways, of course, Crake fits the stereotype of the mad scientist: like figures from Frankenstein and Moreau to Chalfen and Leekie, he is a privileged, brilliant, naïve white man, excessively devoted to rationalism.[11] Crake matter-of-factly tells his friend Jimmy (whom we first meet in his postapocalyptic incarnation, Snowman) that "illness isn't productive" (*Oryx and Crake* 210), just as he informs Ren and Amanda that "illness is a design fault" (*The Year of the Flood* 147). This is almost a direct quotation of Marcus Chalfen's conviction in *White Teeth* that illness is "nothing more than bad logic on the part of the genome" (Smith 260), and its impatience can be heard in many posthumanist voices that raise the battle cry against death.

Without hesitation, Crake joins rationalists like Chalfen in adopting materialist blinders that ignore any foundation for meaning that is not measurable. This perspective conveniently explains away human spirituality by designating God "a cluster of neurons" (*The Year of the Flood* 228), and it motivates Crake to try creating a new species free of any religious yearnings. However, the *MaddAddam* trilogy makes its Frankenstein-Moreau figure into a much more complex character than he might initially seem. The more we understand the context of Crake's upbringing, the less insane he seems. Anticipating how he might be pigeonholed, Atwood gradually reveals compelling reasons behind her bioterrorist's choices, even if they remain horrifying. The trilogy does not become a justification for genocide, but it extends an intensely personal invitation to see how ordinary decisions can cultivate conditions wherein humanity might seem to warrant a mercy killing.

The near-future preapocalyptic world Atwood describes is not yet actual, not quite, and yet it is uncomfortably close. There are disturbing parallels: The de facto power of corporations often exceeds that of nation-states, bypassing the interests of anyone but the wealthy. Citizens are eager to render the line between humanity and other species in bold, when it should be dotted or dashed. And the biotechnologies Atwood describes, from her society's headless protein sources to its many forms of biometric surveillance, are either already realized or imminently plausible. At the same time, suburbia aside, America has not yet been *strictly* divided into corporate "Compounds" and abandoned "pleeblands." Even with the SAT, high school graduates are not *quite* split into "numbers" and "words" people, and rival tech outfits do not *yet* kidnap each other's research scientists on a routine basis. Crake's childhood experiences remain extreme, then, from his mother dying via "a hot bioform that had chewed through her like a solar mower" (*Oryx and Crake* 176) to his father's "accidental" fall off an overpass after objecting to his biotech company's use of its products to spread diseases that it sold others to cure. However credible such episodes might seem, though, the key is that from the earliest age, Crake was surrounded by an ethos of profits above all and a stunning indifference to others' suffering.[12] Perhaps the only real exceptions he witnesses are the God's Gardeners, whose public image might be ridiculous but who still represent the trilogy's most successful resistance to corporate abuses. In this light, it is telling that a disproportionately high proportion of their members survive his plague.

Long after the damage is done, Jimmy makes explicit a question that readers have also been invited to contemplate: "Had [Crake] been a lunatic or an intellectually honourable man who'd thought things through to their logical conclusion? And was there any difference?" (*Oryx and Crake* 343). It is easier to dismiss doubts about Crake's blameworthiness after a first reading of *Oryx and Crake* than it is after finishing the whole trilogy. In the third volume, even some of the scientists whom Crake forces to work with him in the egg-shaped dome called "Paradice" recognize that his choice involves a certain reason. They see that Crake understood how "messed up" the world was, "what with the biosphere being depleted and the temperature skyrocketing," and "in some respects," they credit his casting of humanity as "the greedy, rapacious Conquistadors" and his new creations as "indigenous people" (*MaddAddam* 140). At least in retrospect, readers have even more cause to comprehend the young scientist's perspective. We should notice, for example, that Crake's ultimate project is essentially a large-scale version of a game he used to play with Jimmy, Blood and Roses. As the acts of cultural genocide and artistic achievements pile up, the high-school-aged Crake tells Jimmy, "You can't couple a minimum access to food with an expanding population indefinitely" (*Oryx and Crake* 119–20).[13] Garrett Hardin and fellow lifeboat ethicists might nod their heads at the teenager's logic, which rubs off so fully on Oryx that she tells Jimmy the world's underlying problem is that "there are too many people and that makes the people bad" (322). It is like the reality faced by her anonymous childhood village in Southeast Asia: too many mouths and too little food, so when the rich man showed up seeking more children to "adopt," there was no debate. Similarly, Crake's plague promises a less painful way out, not for an isolated village, but for the entire species. As the level-headed Toby explains Crake's dilemma to his posthuman offspring, "Either most of [humanity] must be cleared away while there [was] still an earth, with trees and flowers and birds and fish and so on, or all must die when there [were] none of those things left" (*MaddAddam* 291). The trilogy neither forgives the scientist his act nor renders it defensible, but Atwood does portray Crake as a severely depressed realist, not a total madman.

In sum, combining death pills, religious performance, and sexual allure yields a nearly complete apocalypse followed by a new genesis. The *MaddAddam* novels are an unrelenting counterattack on the objectification of female bodies, the twenty-first-century endurance of Elmer Gantrys, and the corporate greed that can swallow biotechnological innovations and excrete only nutrition-poor blocks of Soylent Green. Yet the trilogy does not stop with these critiques. Atwood's novels belong near the end of *Editing the Soul* because they reach so fully beyond genetic determinism to also consider the roles of environment and culture, of chance and choice. After the plague, the surviving MaddAddam scientists can be found explicitly discussing epigenetics, remembering that in helping create the Crakers, "the team hadn't been able to eliminate [singing] without producing affectless individuals who never went into heat and didn't last long" (*MaddAddam* 139). Again, music, language, genetics, sexuality, mythology—together or not at all. Even when a narrator like Toby does not fully understand the biology, there are "epigenetic switches to be considered" (217). In short, no matter how much Crake wanted to eliminate the spiritual imagination, he could do so only by deleting the same neurological patterns required for making and grasping metaphors. For all Atwood's amusing sarcasm about traits achieved by various genomic splices, the trilogy rejects the notion that there are merely a few genes to modify and presto! you have a completely different person.

In this sense, Atwood's novels—especially the later two—represent a significant step beyond earlier fiction that enjoyed less scientific cover in challenging genetic determinism. If *White Teeth* ends with its bioengineered FutureMouse scrambling away from a public display case and one of its rebellious protagonists rooting him along—the species-crossing final sentence is, "*Go on my son!* thought Archie"—Atwood's trilogy nearly begins with the announcement that Jimmy's father had just "helped engineer the Methuselah Mouse as part of Operation Immortality" (*Oryx and Crake* 22). Like Smith's novel, Atwood's first *MaddAddam* book reflects its early 2000s publication, a period when the Human Genome Project was just reaching completion. *Oryx and Crake* goes on to include many more references to the exploitation of genetic engineering, as when we watch Jimmy meet "pigoons" developed as cheaper alternatives to "getting yourself cloned for spare parts" (23), a personal insurance scheme like that explored in Michael Marshall Smith's *Spares*, Nancy Farmer's *The House of the Scorpion*, and other cloning tales. Featuring splices of various species' DNA, including human genes tied to neocortex tissue, these giant pigs epitomize the biotech industry's boldness in pushing society toward a capitalistic revival of feudalism. The kinds of executives confronted by Atwood's novel make decisions about whether to rescue an abducted employee via an abstract cost-benefit analysis; they spin their experiences into designer baby startups with names like "Infantade, Foetility, Perfectababe." The result is a commodification of children that makes today's helicopter parenting seem tame, a situation in which kids who "didn't measure up" would be "recycl[ed] . . . for the parts, until at last they got something that fit all their specs—perfect in every way, not only a math whiz but beautiful as the dawn" (250). Atwood's sarcasm is apparent especially in the term *specs*: while defending intellectual curiosity with every breath, her work rejects any biotechnological application that would render a human soul the equivalent of the newest iPhone.

Not only do Atwood's novels reject biology's use as a tool of social control; they also undercut the pseudoscientific narrative on which such practices often rely. There are already hints of this deconstruction built into the first novel—that despite majoring in transgenics, Crake is able to change only part of the picture. Especially in rereading *Oryx and Crake* after its 2009 and 2013 sequels, it becomes doubtful whether the Crakers' mating rituals ever could have been fully genetically programmed. Certainly blue genitalia and new expressions of sexual desire might be imagined as traits dependent on specific genes, but what the later two novels make inescapable is that Crake needed to control much more than genetics to truly decide his creations' future. He needed to lay their cultural groundwork, and for that, he relied on Oryx, who walked among the people in their Egg before the Waterless Flood. Eventually, the experiment takes over, so that after Oryx and Crake disappear, their progeny turn them into divine figures, with Jimmy as prophet. The symbolic behavior and hierarchical self-organization that Crake thought to eliminate from the gene pool quickly reappear just as surely as natural plants and wildlife reclaim urban spaces in this and other postapocalyptic narratives. The sexual behavior presented by the first novel—in which Craker males line up, a female "chooses four flowers, and the sexual ardour of the unsuccessful candidates dissipates immediately, with no hard feelings left" (165)—only works for a time. It ignores the complexity of human desire and the reality that through interaction with previous human beings, if not their own intratribal relationships, the Crakers were bound to discover the individual soul and to begin weighing its desires against rules made for the communal good. Crake attempts to avoid this by abolishing written language—which he considers a gateway drug to abstract, metaphorical thinking—but by the third novel, Blackbeard's curiosity is overwhelming. The genetic dams that Crake erects against personhood and story begin to fail, and the surviving scientists have nothing left to discuss but the epigenetic processes at work. This is the nuance beneath what at first seems a starker determinism, and it is critical for understanding how much more Atwood is doing in these novels than merely lashing out at some corporations. Ultimately she demonstrates the inextricability of contemporary metaphors around genetics, sexuality, and transcendence, and the effects are much greater than most readers immediately realize.

MaddAddam's Secret

Before he was known as the founder of 350.org, a leader who chose arrest at the White House gates rather than complicity with a Keystone XL pipeline deal he regarded as a devastating blow to climate change amelioration efforts, Bill McKibben wrote a book entitled *Enough: Staying Human in an Engineered Age*. Published in 2003, the same year as *Oryx and Crake*, it received Atwood's enthusiastic review. Agreeing with McKibben's suggestion that bioengineering might prove another iteration of the "same old story" wherein humanity poses as divinity, she noted, "It's never had a happy ending. Not so far" (130–31). Focusing on the dangers of a designer baby culture in which human beings would be subject to parental whims for much more than their names, she saw how a genetic arms race could mean that "each new generation of babies will have to have all the latest enhancements—will have to be more intelligent, more beautiful, more disease-free, longer-lived, than the generation before." Children's skills would become outmoded as they were overtaken by the next year's models; people would lose connections to shared ancestors and be cut off from their families. Most worrisomely, class differences would worsen, creating a "GenRich" and a "GenPoor," with the two groups increasingly isolated from each other. "Doubtless 'we' will devise almost impenetrable walls, as in the Zamyatin novel of the same name," Atwood mused, "or 'we' will live in a castle, with 'them'—the serfs and peasants, the dimwits, the mortals—roiling around outside" (138). What she describes, in short, sounds much like the social stratification found in science fiction films such as *Code 46*, *District 9* (2009), and *Elysium* (2013), or for that matter, proposals to build a massive wall across the United States' southern border.

Atwood's response to McKibben's argument maps closely onto her trilogy's fictional world as well. The division between those in castles and those "roiling around outside" is very much the dynamic the *MaddAddam* novels imagine between corporate Compounds and the excluded pleeblands. Such a meeting of the minds should not be surprising, as Atwood has long shared McKibben's concern with climate change denial. Her Booker Prize–winning novel *The Blind Assassin* (2000), for instance, features weather-channel references to global warming and the realization that although people must stop burning so many natural resources, "they won't stop. Greed and hunger lash them on, as usual" (75). By 2006, she was writing satiric pieces like "Chicken Little Goes Too Far," in which the famous character's warnings are rebuked by the teach-the-controversy rhetoric of "Turkey Lurkey," who explains that the sky falling is "due to natural geocyclical causes and is not the result of human activity, and therefore there is nothing we can do about it" (*The Tent* 68). Appropriately, Atwood's meeting with McKibben in 2009 was featured in a documentary about her *The Year of the Flood* book tour, with the two discussing their shared admiration for Henry David Thoreau, both as nature writer and as advocate of civil disobedience.

Given how much her thinking resonates with McKibben's, then, it is striking that her trilogy eventually breaks ranks with McKibben's book about genetic engineering. While sharing his doubts about the altruism of biotechnology companies and their promises for human enhancement, Atwood's genetic metafiction offers a more complex portrait. Undoubtedly, the *MaddAddam* trilogy joins McKibben in eviscerating the corporate greed behind many new biotechnologies, but it indicates that the problem is less with science than with its shortsighted, profit-grabbing applications. Admittedly, on the surface *Oryx and Crake* appears to be yet another story of a mad scientist committing horrors on an unthinkable scale, and there are dozens of passages in which readers are invited to recognize the commodification of glow-in-the-dark animals and other transgenic experiments.[14] Beneath the eviscerations, though, there is also a recognition that real good might be done through biomedical advances. As Atwood says in a preface to *Gulliver's Travels*, "We are not only what we do, we are also what we imagine. Perhaps, by imagining mad scientists and then letting them do their worst within the boundaries of our fictions, we hope to keep the real ones sane" (211). This is reason to attend to not only the trilogy's many satirical elements but also its overarching effect as a speech act. The novels are not shy about their many targets, but they also convey subtler ideals: a form of spirituality concentrated on action rather than doctrine, a love that offers mercy where it is undeserved but refuses sentimentality or self-righteousness, and an integration of scientific knowledge, emotional maturity, and theological speculation that embraces rather than flees uncertainty.

One of the best readings of these novels to date comes from Gerry Canavan, who makes a compelling argument for their status as satire and allegory. Writing before the third novel's appearance, Canavan dismisses the notion that either the Crakers or the God's Gardeners represent a serious proposal for dealing with climate change or biotechnological corporatism. As he explains, "The Crakers allegorize the radical transformation of both society and subjectivity that will be necessary in order to save the planet—showing us how very difficult the project will be, and giving us a sideways, funhouse-mirror, only-kidding glimpse at the kinds of revolutionary changes that will be required to make the future better than the present" ("Hope, but Not for Us" 152). This recalls various metafictional moments in *Y: The Last Man*, as Canavan rightly emphasizes Atwood's Swiftian epigraph for *Oryx and Crake* and the satirical lens by which she has often addressed cultural extremes, whether they rely on superficially religious or pseudoscientific rhetoric. *MaddAddam* aside, his essay convincingly shows that the story to that point idealizes neither "eco-religious separatists" nor posthuman simpletons and that it should not be misunderstood "as a blueprint for utopia, nor as a Bible for the world to come." Instead, the first two novels are "asserting through allegory the urgent necessity of radically changing our social relations and anti-ecological lifestyles—of choosing to make a better social world before it is too late for the natural one" (155). In short, Canavan makes an excellent argument for reading *Oryx and Crake* and *The Year of the Flood* as genetic fantasies with pressing environmental applications.

However, I also think Atwood's third novel changes the game significantly.[15] Demanding a rereading of the whole trilogy, it epitomizes genetic fantasy's capacity to mutate into metafiction. My claim is not just that Atwood complicates each novel with the subsequent one but that there were always layers upon layers in them that only became visible with the trilogy's completion. Those familiar with Atwood's larger oeuvre will not find this shocking: she is a master of structural innovation, but unlike some of her postmodern brethren, she does not necessarily flaunt it. Think, for instance, of her self-revising short story "Happy Endings," the famous epilogue of *The Handmaid's Tale*, or more recently, *The Blind Assassin*. The latter features a novel within the novel, which itself includes an internal science fiction tale, so that much of the work's poignancy emerges only as the reader pieces together the three narratives' relationships. With the *MaddAddam* trilogy, the tripartite nesting doll is easier to overlook, and it works in reverse—from the inside out instead of the outside in. Providing the inner "doll" first, *Oryx and Crake* features many of the up-close glimpses of what occurs immediately before, during, and after the plague's release. *The Year of the Flood* then moves outward to focus more heavily on the years leading up to Crake's decision and the alternative perspective of the

God's Gardeners. Finally, *MaddAddam* is the external nesting doll that we view last, and its details about the plague's origins and its denouement dramatically reframe the two previous works. It is not that it cancels out the preceding satire or allegory—it has plenty of moments that extend the fun, as when Zeb feeds on flank meat from his would-be assassin, "Chuck," to survive the Alaskan wilderness. It maintains Atwood's delightfully impish humor and expands her ecological purview, but it also drives home additional insights about the ties among scientism, religious absolutism, sexism, and corporatism. My argument is that fully understanding Atwood's combination of satire and metafiction depends on appreciating a startling trick of perspective that she holds up her sleeve until the final novel.

The surprise is that despite all appearances, Crake has not been the mad scientist behind everything that he seemed, that Atwood is pointing her finger well beyond the geneticists, and that, in fact, nearly everyone is implicated. Admittedly, the first novel sets up readers to interpret Crake as its prime mover, and the second does little to undercut that assumption. The genius of *MaddAddam*, though, is in revealing not just that Crake had help but how he served as still another character's tool. Most readers of the third novel will grasp that Crake forced the MaddAddam scientists to join his project and that he had earlier connections to the God's Gardeners than were previously obvious, but the real bombshell is most impressive for the sheer fact that one can miss it entirely, much like the palindrome of the third novel's title. As the remainder of this chapter argues, Atwood's trilogy is transformed once readers realize that the man behind the curtain was less often Crake than Adam One. Once this becomes clear, the novels' satirical overtones become doors to a more urgent manifestation of genetic metafiction. Without deleting the humor, there is suddenly far greater significance in moments like Zeb's reflection, "Best thing old Slaight taught me was misdirection. Make them look at something else, away from what you're really doing, and you can get away with a lot" (*MaddAddam* 171). What Atwood is "really doing" in this trilogy is not just entertainingly warning her species about the dangers of its current path but showing how closely related are contemporary concepts of genetics and cultural ideals, how below them lie foundational myths about biology and spirituality, the body and the soul, human beings and other animals.

Before reexamining Adam One through the lens of *MaddAddam*, it helps to recall the man who leads the God's Gardeners in *The Year of the Flood*.[16] One of the strangest, most fully realized, and least stereotypical new religious groups to ever grace the pages of fiction, the Gardeners provoke readers to ask of Atwood, *Is she serious?* Or better, *How exactly is she serious?* This ecoreligious cult is over-the-top cheesy, yet one quickly begins rooting for its wild characters. Many readers are likely to join Canavan in focusing on the sly humor with which Atwood approaches the story. Even in the first novel, there are guffaws aplenty, like the "botched experiments" during the creation of the Crakers, wherein "one of the trial batch of kids had manifested a tendency to sprout long whiskers and scramble up the curtains; a couple of the others had vocal-expression impediments; one of them had been limited to nouns, verbs, and roaring" (*Oryx and Crake* 156). Crake's world features bath towels made from algae that exhibit a bad habit of "puffing up like rectangular marshmallows and inching across the bathroom floor" (202). There are variations on Viagra with disturbing results, even before Crake decides to repurpose the BlyssPluss Pill: "A couple of the test subjects had literally fucked themselves to death, several had assaulted old ladies and household pets, and there had been a few unfortunate cases of priapism and split dicks" (295). And stepping back far enough, it grows apparent that the young man has constructed a species that quite literally holds daily pissing contests and eats its own feces. Not surprisingly, then, *The Year of the Flood* expands the satire with its tales of SecretBurgers (guess what's in the "meat" today!) and Petrobaptists (Jesus gave Peter the keys to the kingdom, his name means "rock," and that's where we get oil . . . so watch out if Halliburton buys a church in your neighborhood).

The God's Gardeners are the centerpiece of the second novel's seeming absurdities. Atwood's adaptations of classic English hymns are simultaneously touching and hilarious in their sentimentality (it's hard to beat "Oh Sing We Now the Holy Weeds"), especially with Zeb playing a (partially) reformed Miltonic Satan in the background, the amusingly devious character who smirks at all forms of excessive piety. When Adam One suggests "they should take a moment of silence and put Light around Burt in their hearts"—he had been running an illegal marijuana gro-op right under his fellow Gardeners' noses, the exposure of which cost many of their homes—Zeb responds that "they'd already put so much light around him the guy was probably burning like a suicide bomber in a fried-chicken franchise" (*The Year of the Flood* 191). There is more humor where that came from, whether via Pilar and Toby talking to their bees or Adam One suggesting in a sermon that Jesus called two fishermen as disciples in hopes of conserving Lake Galilee's fish population. Yet all the silliness can obscure the fact that Atwood has hidden the deepest source of her genocide in plain sight. As Toby later illustrates with her duck furzoot costume, the best camouflage is to be taken as a joke. By running such a humble operation with so many pathetic characters, Adam One shields himself from the gaze of not just the CorpSeCorps (the police now answer directly to corporations) but also the reader. Like his creator, he is a master of disguise, a new kind of preacher who is ready to "cheerfully admit the absurdity—from a materialist view—of every Spiritual truth we profess" (196). As it turns out, this is not just a wise stratagem but a genuine confession of his ulterior purposes.[17]

The God's Gardeners, that is, are a worthy mission in and of themselves but also a means to an end that lies only on the other side of apocalypse. Adam One does not preach Daniel or Revelation with an eye to the rapture but the Waterless Flood to those he is preparing for another kind of global transformation. What we learn in Atwood's third novel is that Adam is even wilier than Crake. He maneuvers Zeb on the chessboard just as fully as Crake does Jimmy (and sometimes Zeb), and then he finally nudges Crake himself into place. A true grandmaster, Adam puts the most virulent of bioweapons in an angry teenager's hands even as he prepares his community for the effects. This becomes evident only when we reread what Atwood has previously told us of the God's Gardeners leader in light of the additional backstory provided by *MaddAddam*. With a light touch that refuses to render Adam the kind of self-deluded religious charlatan we find in Carl Sagan's *Contact* (1985), Sinclair Lewis's *Elmer Gantry* (1927), or other novelistic rebukes to narrow-minded religion, Atwood asks readers to wrestle with Crake's impossible choice rather than shrugging it off as inherently insane. As in the novels and essays of Walker Percy, the implication is that humanity might only be able to truly live after giving serious consideration to self-destruction.

Those who look back carefully at the first two *MaddAddam* novels in light of the third will find much more occurring below the plot's surface than is directly conveyed. Even before the third novel's publication, for instance, Canavan sensed how "the implication of [*The Year of the Flood*] is that Crake has secretly been working on his virus for nearly his entire life, originally testing it on his own mother in retribution for her betrayal of his father; the symptoms of the world-ending supervirus match quite closely the symptoms of the unknown infection that killed her" (156n4). More of Adam's backstory appears in the final novel, especially via his juxtaposition against Zeb. We hear how Adam "used good things as a front for his bad things" (114), that "Adam's knowledge of banking surprised [Zeb], but then Adam's knowledge of a lot of things surprised him. It was hard to know what Adam knew" (122). Most critically, Adam purposefully arranges Zeb's murder of the man who raised them, the Rev, yet without providing any explicit instruction. He simply understands his younger brother's character, gives him the necessary tools (the pills hidden in the serpent-woman saltshaker), and puts him in an opportune position on the chessboard. As it turns out, this is exactly the strategy he uses years later with Crake. Adam never *tells* Crake to launch a virus that will destroy almost all of humanity. He simply recognizes the young man's potential, feeds his hunger for knowledge, and ensures that Crake eventually receives the biochemical means.

Zeb nearly says as much in the third novel when he explains to Toby how Crake came to possess the remaining three pills, those not used to turn Zeb's fraud of a father figure into pink goo. Acting with Adam's full approval, Pilar willed her chess set, including the pills lodged within the bishop, to Crake. Toby asks, "Do you think Pilar knew what use he'd make of those microbes or viruses or whatever they were? Eventually?" Then she remembers Pilar's face and its "hard resolve," and Zeb responds, "Let's put it this way. All the real Gardeners believed the human race was overdue for a population crash. It would happen anyway, and maybe sooner was better" (*MaddAddam* 330). This becomes overwhelming evidence that Adam One actively enabled the Waterless Flood and even knew how it would spread. Recall that well beforehand, he admonishes his followers, "We God's Gardeners are a plural Noah: we too have been called, we too forewarned.... We must be ready for ... the Waterless Flood, which will be carried on the wings of God's dark Angels that fly by night, and in airplanes and helicopters and bullet trains, and on transport trucks and other such conveyances" (*The Year of the Flood* 91). It is not just another strange hygienic obsession that leads him to insist the Gardeners wash their hands "seven times a day at least, and after every encounter with a stranger." Particularly in exhorting, "It is never too early to practise this essential precaution" (92), he betrays an awareness of the Waterless Flood's future timing and its means of dissemination. He knows the specific means by which Oryx will distribute the virus and the devastation it will cause, and when it finally arrives, he does not despair but exults. His last recorded sermon is therefore heavier on celebration than lamentation: "But how privileged we are to witness these first precious moments of Rebirth! How much clearer the air is, now that man-made pollution has ceased!" (371).

Adam's sermon also indicates that while he might have been manipulating Crake more than it seemed, they share a commitment to protecting human(like) life on Earth. They are the ultimate ecoterrorists, yet Atwood makes it hard not to like them, at least a little. What Adam preaches—humanity's failure to "live the Animal life in all simplicity" (52)—is exactly what Crake attempts to recover via the Crakers. By hiding behind his brother, Zeb, the outwardly violent and uproarious God's Gardener whose nickname is "the Mad Adam," the trilogy's ultimate puppeteer is able to operate without suspicion. In retrospect, of course, there are plenty of hints. In *The Year of the Flood*, for instance, one learns how Toby "had come, over time, to realize it would be a mistake to underestimate [Adam One]. . . . Toby felt she would never encounter anyone as strong in purpose" (97). This is key to his success: unlike most Elmer Gantrys, Adam One is a true believer in the sermons he preaches, even regretting the necessity of occasional white lies. He recognizes the limitations of theology and of language more generally—humanity's inability to say anything incontrovertibly true—so while remaining masked beneath a seemingly infinite supply of corny sermons, he fears that "it is not this Earth that is to be demolished: it is the Human Species. Perhaps God will create another, more compassionate race to take our place" (424). It is highly likely that he not only orchestrates Crake's genocide but knows the young scientist's plan for repopulating the Earth.

On grasping Adam One's expanded role in this "Paradice" Lost retelling, some readers might only expand their disgust with the trilogy's mad scientist to include his religious freak of a brother. Events suggest that these stereotypes are merely two sides of the same coin, but Atwood is suggesting a good deal more about both biology and religion here. Compare the Adam One emerging from this trilogy to another cult leader in recent fiction, the unnamed prophet of Emily St. John Mandel's *Station Eleven* (2014), itself a remarkable postplague tale. Mandel's figure bears many superficial similarities to Adam One, but he is a far less finely drawn character. One cannot imagine an actor playing the role and being taken seriously as a source of anything good; his tone is even more sickly sweet, especially when he asks his people to "consider the perfection of the virus" that killed so many: "There were no more statisticians by then, my angels, but shall we say ninety-nine point ninety-nine percent? One person remaining out of every two hundred fifty, three hundred? I submit, my beloved people, that such a perfect agent of death could only be divine. For we have read of such a cleansing of the earth, have we not?" The distinction from Adam One's saccharine meditations might not be enormous, but it is critical: Adam One celebrates the cleansing effects of his own Waterless Flood but never *the means*, nor does he suggest that it was divinely approved. Moreover, he remains humble despite predicting the disaster to come, never ascribing his people's survival to his or their worthiness. By contrast, *Station Eleven*'s cult leader goes on, "The flu, the great cleansing that we suffered twenty years ago, that flu was our flood. The light we carry within us is the ark that carried Noah and his people over the face of the terrible waters, and I submit that we were saved not only to bring the light, to spread the light, but to *be* the light. We were saved because we *are* the light. We are the pure" (60, emphasis in original). Such naïve self-righteousness never escapes Adam One's lips; perhaps he read Zadie Smith too carefully.

Put another way, Adam One is just as manipulative as Mandel's prophet, but with the enormous difference that he genuinely believes in the necessity of his actions. Even as a child, his "blue-eyed lying" was capable of "mak[ing] just about anyone except Zeb think he was innocent as an egg unlaid" (*MaddAddam* 116). What separates Adam One from most megalomaniacs and grandiose liars is that he stays aware of his fictions even as he creates them. As Zeb reflects, his childhood sibling decided "to do the Rev thing himself, but do it right—everything the Rev had pretended to be, he would be in reality" (333). This does not mean exploiting the credulous but rather preparing his flock for the new world he knows must come. Rather than believing he can save the world completely, Adam One knows that his project is "not perfect. He wouldn't claim that. More like a reboot" (334). It is an effort to begin anew by harnessing the power of fiction, reestablishing the close ties between story and history that are lost when fundamentalist religion and militant scientism go to war. Adam One's greatest tool, in other words, is also Atwood's: a constant metafictional awareness that the same truth can yield to multiple interpretive approaches.[18]

In examining the whole of the *MaddAddam* trilogy, Adam One's concept of the relationship between religion and literature is also revealing. Contra the vast majority of cults in contemporary fiction, the God's Gardeners do not adhere to a rigid, literalist, hyperauthoritative hermeneutic. The very name they give their holy books—which might include texts from beyond Judaism and Christianity—is telling: "the Human Words of God." These do not constitute a directly-applicable-to-all-circumstances divine rulebook, nor are they easily shunted aside as wishful sayings or just-so stories about the inexplicable. Rather than acting like a magic wand, scripture operates for Adam's brethren as both holy and fully human, inspired but unfinished. One major effect is relatively little perceived conflict between religion and science. As Adam One explains,

> The Human Words of God speak of the Creation in terms that could be understood by the men of old. There is no talk of galaxies or genes, for such terms would have

confused them greatly! But must we therefore take as scientific fact the story that the world was created in six days, thus making a nonsense of observable data? God cannot be held to the narrowness of literal and materialistic interpretations, nor measured by Human measurements, for His days are eons, and a thousand ages of our time are like an evening to Him. Unlike some other religions, we have never felt it served a higher purpose to lie to children about geology. (*The Year of the Flood* 11)

Adam One's position, in other words, avoids the narrow-minded assumptions common to militant forms of both theism and atheism, and it also reaches beyond relatively simple truces that would pretend science and theology never intersect.

Rather than banishing the fighters to their corners, Atwood's most influential character seeks potential integrations of religious and scientific insights, yet without forcing them. He is too comfortable with mystery, preferring better questions over final answers. As Toby recalls, "Some will tell you Love is merely chemical, my Friends, said Adam One. Of course it is chemical: where would any of us be without chemistry? But Science is merely one way of describing the world. Another way of describing it would be to say: where would any of us be without Love?" (359). For Adam One, even with all his sappiness, the two languages interpret one another. Or as Toby eventually learns, "just because a sensory impression may be said to be 'caused' by an ingested mix of psychoactive substances does not mean it is an illusion" (*MaddAddam* 227). One way to experience reality is under the microscope; another is via relationships within communities and across differences. Healthy binocular vision requires that these lenses be carefully aligned. At the same time, for some questions, they must be allowed to operate independently.

In this book's terms, what Atwood puts forward is neither a religious nor a secular viewpoint but a postsecular one. Thus it should not be surprising that the third book in the trilogy finally mutates into a kind of scripture itself, at least according to the God's Gardeners definition. The oral narratives provided first by Snowman-the-Jimmy and then Toby are eventually undertaken by Blackbeard, the Craker who is on stage most extensively. Taught to read and write by Toby, his voice takes over *MaddAddam*'s narration in its final pages: "This is the Book, these are the Pages, here is the Writing" (385). This is possible because the stories told by the keeper of the red cap have always been open-ended, as if Atwood is attempting to remind us that other scriptures might also have been conceived not as finalized documents to be venerated by generations to come but as the gradually unfolding prose and poetry of peoples in the process of becoming. As such, the nightly fireside stories told by Snowman-the-Jimmy, Toby, and then Blackbeard do not represent a closed canon, nor does the written version we begin to sample. In the trilogy's final pages, Blackbeard conveys Toby's instructions about the Book, most critically the instruction that each newly literate Craker should make another copy of the Book, and "at the end of the Book we should put some other pages, and attach them to the Book, and write down the things that might happen after Toby was gone" (386). In this sense, each version will pirate the one before it, but it will be a gentle, voluntary piracy, much like the demeanor of Blackbeard. This explains his narration of "The Story of Toby," the final entry in the text readers encounter as *MaddAddam*—which cannot be the Book's actual ending. Atwood's proposal is that scriptures are stories about the things that matter most, and so long as there are lives to narrate, so long as there is DNA to mutate, so long as there is epigenetics to thicken the plot, the Book should go on.

Moreover, Atwood is saying something via the *MaddAddam* trilogy about *all* stories, not just the ones adopted as orthodox or especially noteworthy. As genetic metafiction, her trilogy points to the inherited nature of human narratives, and it urges us to come to grips with the possibilities of fictional ambiguity in particular. Among their many roles, the Crakers serve as funhouse mirrors for our own befuddlement about how to align the intuitions of words with the relative objectivity of numbers. At first, these transhuman characters stumble even with the fantasy involved in pictures, but gradually they realize how "*not real can tell us about real*" (*Oryx and Crake* 102). In this sense, they advance well beyond the Painball imbeciles, who have a great deal more experience but still "didn't understand make-believe" (*The Year of the Flood* 130). By contrast, Blackbeard demonstrates great imagination as he matures across the third novel.[19] After discovering the "smelly bones" left behind by his gods, he shifts from literalism into metaphor. Initially he objects, "Oryx and Crake must be beautiful! Like the stories!" (356). Not much later, though, we find him taking up the prophet mantle himself to extend those very stories. To do so, he relies on the early stages of a Platonic dualism, suggesting that "the Egg wasn't the real Egg" and that "Oryx and Crake had different forms now, not dead ones" (360). This is the beginning of philosophy, and one might expect that it will also evolve as the Book expands.

One can hear certain voices that would respond, "Really? We're going to do all *that* again?" The Ray Kurzweils of the world, for instance, express hopes for new forms of being that escape the body far more thoroughly than Atwood's dick-waving, orgy-loving Crakers seem to manage. But in the *MaddAddam* trilogy, she achieves something beyond the kind of end-of-humanity apocalypse typified by Arthur C. Clarke's *Childhood's End* (1953). Rather than trying to wipe clean the slate of history, impatiently pushing for humanity to become something wholly different, she embraces a more gradual evolution. This still involves fairly radical bioengineering, to be sure, but as Crake explains to Jimmy, "You've got to work with what's on the table." If humanity is prone to war and "misplaced sexual energy," then it should do more than tweak genes associated with violence and inhibition (*Oryx and Crake* 293). It should confront overpopulation, ecological collapse, the problem of nutrition. The same applies to building cultural mythologies and values, as Toby and Snowman-the-Jimmy discuss much later. The Crakers wanted to know about their origins, so their half-drunken Moses makes up a story. He might regret many of its details, Toby observes, "'but that's the story we've got,' says Toby. 'So we have to work with it'" (*MaddAddam* 265). She once queried Adam One about why it was necessary to have "God in the story," and he explained, "This belief bias of ours must confer an evolutionary advantage. The strictly materialist view—that we're an experiment animal protein has been doing on itself—is far too harsh and lonely for most, and leads to nihilism. That being the case, we need to push popular sentiment in a biosphere-friendly direction by pointing out the hazards of annoying God by a violation of His trust in our stewardship" (*The Year of the Flood* 241). Beneath the piety, this is Adam One's pragmatism at work, and it receives significant pride of place in Atwood's trilogy. Toby recalls hearing much the same message from the clearest wise woman in the novels: "People need such stories, Pilar said once, because however dark, a darkness with voices in it is better than a silent void" (*MaddAddam* 154). The combined weight of these reflections on the problem of meaning is ultimately very significant: however much individuals might sometimes wish to free themselves from subjectivity, emotion, or even their longing for transcendence, these "limitations" lie at the core of the soul. There can be no true freedom without them.

This paradox brings us back around to the question of how Atwood is serious with this trilogy, particularly in her depiction of the God's Gardeners. My position is that she is that rare artistic creature who is simultaneously capable of hilarious satire and zealous urgency, a contemporary Shakespeare who writes fewer tragedies or comedies than tragicomedies. And she too sponsors dramatic performances: in taking a book tour across the United

Kingdom, the United States, and Canada for *The Year of the Flood*, she participated in some forty presentations in eighty days, mostly in relatively liberal/mainline Protestant churches or with their assistance. In each case, she sent ahead a script for several actors and musicians, and a wide variety of amateur and professional groups were assembled to perform scenes from the novel and sing the God's Gardeners hymns. Atwood provided Orville Stoeber the rights to set these to music and later to release them as a kind of "book soundtrack," and she served as narrator/reader for the events, which often featured local clergy in the role of Adam One. In other words, in the time between the publications of the trilogy's second and third books, her regular interest in publishing via alternative formats took one of its strangest turns ever. She shrunk the distance between herself and her narrator to the thinnest possible gap, inviting predominantly Christian artists to incarnate the God's Gardener characters she had created on the page. Especially when performed at the front of churches, the effect was closely akin to a sermon illustration (fig. 17).[20]

What do these events say about Atwood's novels? Here one must remember, as a chapter in *MaddAddam* begins, "there's the story, then there's the real story, then there's the story of how the story came to be told. Then there's what you leave out of the story. Which is part of the story too" (56). When *MaddAddam* was still yet to be published, Atwood had left a great deal out of the story—perhaps elements she was still recognizing herself. After all, how many of these clergy would have participated in her book tour performances had they known their character enables global genocide? Adam's fall, however, is ultimately very fortunate. Atwood's promotional work for *The Year of the Flood* reached beyond common ideological barriers and exhorted Christian communities in particular to rethink their theologies of the Earth. She used metaphor, the central means by which we humans have long approached the numinous. Toby says of this inexact, evocative capacity of language, "It wasn't true or not true. It was in between. It was a way of telling about a feeling you might have" (*MaddAddam* 109). My argument is essentially an extension of this logic: the *MaddAddam* trilogy concerns not what Atwood is saying biotechnology *will* make of our culture but what it and its owners *could* make, what such forces are *already* making. The story is a way of telling about a feeling she has, a route beyond one-to-one referentiality to a hermeneutic space in which truth cannot be absolutely measured or proved yet is all the more meaningful for that very reason. This is what science needs literary and religious speculations

Fig. 17 Margaret Atwood (far right) reads during her book tour's first performance of scenes and hymns from *The Year of the Flood* (2010).

alike to offer: a way of reaching beyond the search for simple objectivity that sometimes must reign supreme in the laboratory but that, left to itself, risks producing only carbon-copy clones—or at least the violence that ensues when corporatism demands complete conformity and human beings fail to provide it.

On the eight-year anniversary of the 9/11 disasters, Atwood sat for a televised conversation in Canada with three other notables, including a clergyman and Richard Dawkins. That evening's *Newsnight Review* topic was not terrorism but "Darwin on Stage and Screen," and the clashes to come were predictable. When two of the panelists suggested that natural selection might be the primary means of God's creative work, Dawkins scoffed, "It is astonishing the way somebody can let their emotions rule over factual belief. I mean, it doesn't matter what your emotions tell you: if it's true, it's true, and you better live with it." This was only an early salvo in the exchange, but Atwood's chuckling response was telling: "But that's not how people are!" Channeling Adam One, she later explained, "There's only a few Richard Dawkinses, but there's a ton more people who would get behind an attempt to save the planet if they believed it was their duty. And fundamentalism in the United States is split into two groups. One group is going green; they've even put out this green Bible, it's got the green bits in green. And the other ones are the rapturists who think, well, it's all going to be burned up anyway except us, and then we're going to get a new one, sort of the ultimate consumerism."[21] This is Atwood's pragmatism flying in the face of Dawkins's militancy. From his perspective, she is another disappointing "accommodationist," a sellout like biologist and atheist E. O. Wilson, who appealed to Christians in his 2006 book *The Creation* "to save life on Earth." From her perspective, Wilson is a saint among the God's Gardeners, one of many nonreligious but deeply moral figures they venerate. Atwood believes this kind of adulation is an "evolved adaptation," and she accepts it as well as the need to partner with willing religious people in saving the planet. As she puts it, "You can't say to people, 'Thou shalt not have any religion,' because it's more or less wired in, so it's not a question of whether they will have it [religion] or not, but what kind they will have." When Dawkins objects to the phrase "wired in," she clarifies, "OK, a tendency, an inherited tendency towards . . . [something] which could in fact be channeled into fervent stamp collecting or something like that." Or, she puckishly adds, "Maybe being a biologist." Dawkins, she suggests, also has his ideology, and it is part of the problem, not the solution.

In other words, Atwood insists that the most scientific of behavior can be undertaken religiously, just as religious behavior can be approached as if its real meaning were statistically measurable. Neither of these represents the kind of cross-disciplinary, multidimensional thinking that she is encouraging. Rather than assimilating one of these fields within the other, and rather than pretending they have no concerns in common, the goal should be to pursue related questions from the multiple angles provided by very distinct methodologies. Such projects can deeply inform one another, especially in eliminating demonstrably wrong answers (e.g., the idea that the Earth is flat or six thousand years old, on the one hand, or that all human beings will experience transcendent awe in the same manner, on the other). But they must not be confused for one another: we are on dangerous ground whether we assume that all forms of wisdom are accessible via scientific investigation or that all forms of knowledge require only a scriptural proof text. We are animals, yet more than the sum of our body parts, to both our benefit and our harm. Listen to Atwood combine linguistic and biological inheritance once more in her conversation with Dawkins, in this case after being asked about anthropomorphism:

> Animals don't have the past perfect tense or the future perfect tense. It's not that they don't have memory, and it's not that they can't anticipate events. They remember that you came back yesterday and they hope that you're going to come back at 5:00

and give them their food. They anticipate you. But no dog, as far as we know, has ever said, "Where do dogs come from in the beginning?" or "Where will I, Rover, personally go when I die?" As far as we know, they're not saying that. But language gives us the ability to project infinitely backwards and infinitely forwards, and that's our problem.

That's our blessing and curse, Atwood says—our ability to extrapolate so far beyond the immediate present. When we refuse to acknowledge the limits of our reach, eventually the ladder topples. Notice the parallel with Toby's summary of Adam One's teaching about the origins of humanity and its fall from grace: "Then they fell from instinct into reason, and thus into technology; from simple signals into complex grammar, and thus into humanity; from firelessness into fire, and thence into weaponry; and from seasonal mating into an incessant sexual twitching. Then they fell from a joyous life in the moment into the anxious contemplation of the vanished past and the distant future" (*The Year of the Flood* 188). Seeking divinity, humans reach for the unknown. The sin is not ambition or curiosity but letting the remote past or future obscure the present.

Story matters for much more than its capacity to be turned into so-called life lessons, but perhaps there is a humble rule of thumb here for evaluating biotechnological futures. The question to ask is this: Does a new tool cultivate a richer life in the present and the foreseeable future, or does it cultivate distraction from or even denial of the reality of individual transience? This question might provide new perspective on the problem of "status quo bias" raised by bioethicist Ronald M. Green. The operative question might be, Does this genetic test, this gene therapy, this experiment in gene editing hold out the promise of better lives for all concerned in the here and now, or does this effort reflect unwillingness to face actual circumstances? And then, might the same tool mean opposite things for different users, or even the same user in different times and circumstances? Barring the widespread success of cryogenic processes like those purchased by the rich in Don DeLillo's *Zero K* (2016), the reality is that everyone has only a limited number of breaths, irrespective of whether they believe some form of personhood lies beyond the veil of the tangible. The world of the early twenty-first century is vastly different from that of our grandparents, just as our grandchildren might take for granted realities we can barely anticipate. Our stories, though, serve as critical means of stretching backward and forward in time, limbering us for the twists and turns of life's journeys. At times, it is possible to feel like

Jimmy channeling Hamlet, "whang[ing] up against the Great Indifference of the Universe" (*Oryx and Crake* 260); other times, one is left as high as an Eve on a mushroom-fueled vigil. Either way, Atwood's self-reflexivity places her trilogy among literature's most incisive analyses of humanity's capacity to redesign itself. The novels examined in chapter 5 find metafiction similarly reshaping genetic realism.

5.
The Predisposed Agency of Genetics and Fiction

Near the end of Margaret Atwood's *Oryx and Crake*, Jimmy awakens to his last day alone with the Crakers. He has been the story's hapless Moses, chosen by his old friend to lead humanity 2.0 out of its Egg and into the wilderness of a far less populated world. The biblical Moses rebelled at times against his responsibility, but Jimmy finds the prophetic life even more difficult to tolerate. Unlike the ancient Israelites, the Crakers are a cheerful, easily contented people, so he must do all the grumbling himself. Ineligible for their orgies, he finds temporary solace in the liquor cabinets of abandoned suburbia, but when the supplies run dry and he learns other humans are nearby, he decides to follow the footprints in the sand. First, though, the would-be Crusoe looks out from his tree-limb sanctuary: "On the eastern horizon there's a grayish haze, lit now with a rosy, deadly glow. Strange how that colour still seems tender. He gazes at it with rapture; there is no other word for it. *Rapture*. The heart seized, carried away, as if by some large bird of prey. After everything that's happened, how can the world still be so beautiful? Because it is. From the offshore towers come the avian shrieks and cries that sound like nothing human" (371, emphasis in original). Jimmy is wrested into the present by a sunrise and birdcalls that are at once sublime and ordinary. The moment's "deadly glow" heralds neither Revelation's rapture nor Romantic idealism about "Nature" but the eternal now, the alien splendor of the postapocalyptic everyday. Finding new solidarity with the nonhuman, Jimmy descends from his rustic bedroom, pees on his grasshopper friends, and prepares to rejoin the species.

At the end of chapter 4, we saw how Atwood targets humanity's obsession with the abstractions of an infinite past and an infinite future. In the *MaddAddam* novels, as in her conversation with Richard Dawkins, she instead draws attention to the moments in which we actually live. While her tone is often satiric, exposing the consequences of consumerism and its distractions, she also gestures toward an alternative. Jimmy's sunrise vision is one example: even if only briefly, its otherworldly shrieks draw him back into "joyous life in the moment," letting him renounce "the anxious contemplation of the vanished past and the distant future" (*The Year of the Flood* 188). Compare the scene with a parallel moment in Richard Powers's 2009 novel *Generosity*. Two of the main characters are finally onstage together: Powers's magnetic, entrepreneurial synthetic biologist, Thomas Kurton, and the girl-next-door host of a popular science edutainment show, Tonia Schiff. They have just finished recording an interview at his second home in Maine and the technicians have left for the hotel. (One of them just won a bet that she would be "delayed" by their subject.)

Tonia predicted this development herself, and it feels "lovely" when he first touches her arm, but she reflects, "Dopamine, serotonin, oxytocin. Does knowing the chemistry change anything? How long ago did she discover that *lovely* was a chemical trick?" (176). Seated on Kurton's wraparound porch as he prepares dinner, she realizes that accepting his invitation was "not even a decision" but "just large molecules, passing their oldest signals back and forth across the infinite synapse gap." And then everything changes: "A noise comes from up in the woods. She can't tell if it's a mammal, bird, or something stranger. A throat considerably smaller than hers, but monstrous compared to the rest of creation, moans in spectral restlessness. She waits until the sound returns. It's a call from back long before contentment and agitation parted ways." The effect of this alien cry is even greater than in Atwood, taking Tonia beyond Jimmy's rapture to the shocked perception, "*I have no center*" (179, emphasis in original). Powers employs a favorite refrain here: Tonia realizes that "she's disappeared into playing herself" (179–80). The result is that she finally makes a real choice, abandoning the impromptu date and wherever it might lead in favor of recovering the storytelling integrity that is her true vocation.

This final chapter of *Editing the Soul* focuses on a convergence of the soul, genetics, and a larger sense of ecological integration via two of Powers's most recent novels. While attentive to reverberations across his previous fiction, interviews, and essays, I demonstrate how *Generosity* (2009) and *Orfeo* (2014) epitomize the fictional evolution assessed across this book's chapters. Admittedly, my goals include attracting potential readers to these fictional labyrinths, and for those who have already taken a tour, inspiring richer rereading. My primary aim is not merely to analyze Powers's individual achievements, though, but to indicate precisely how they build bridges between worlds that often seem estranged.

In demonstrating the capacity of genetic realism to mutate into a profoundly self-aware genetic metafiction, these novels reveal shared epistemological foundations beneath science, religion, music, and literature. Indeed, it is critical that the eureka moments in which Jimmy is struck by the world's enduring beauty and in which Tonia rediscovers her freedom to choose are responses to the invisible cries of seemingly alien beings. Like Atwood, Powers calls us back to our creaturely natures without in any way denigrating the human. There is no less mystery, wonder, or awe in being earthly; in fact, there is a new glory in the physical world. This claim might seem far afield from an argument about genes in fiction, but more than any novelist I know—except perhaps Kim Stanley Robinson, in his own way—Powers's oeuvre holds together global ecology and local microbiology. Reaching beyond the superficial, spectacular rhetoric of much popular science, a dynamic vision of the genome serves here as an opening for the soul rather than a means to its assimilation or reduction. Here the material is wondrous and the spiritual is inescapably embodied. Fiction emerges as a living pattern very much like a genome—easily mutated but persistently self-replicating, relying on beginnings and endings that are ultimately a matter of perspective.

The Glory of the Ordinary

Given its reach, it should be no surprise to find Powers's fiction extolled by a fellow novelist, but when Atwood reviewed *The Echo Maker*, Powers's 2006 National Book Award–winning mystery about cranes, cognition, and global capitalism's costs for rural culture, she explained why many are yet to fully appreciate his work:

> If Powers were an American writer of the nineteenth century, which writer would he be? He'd probably be the Herman Melville of *Moby-Dick*. His picture is that big. *Moby-Dick* sank like a stone when it first came out: it had to wait almost a century before its true importance was recognized. Given Powers's previous interest in devices like time capsules, I'd hazard that he has the long view in mind: open him up in a hundred years, and there, laid out before you in novel after novel, will be the preoccupations and obsessions and speech patterns and jokes and gruesome mistakes and eating habits and illusions and stupidities and loves and hates and guilts of his own time. All novels are time capsules, but Powers's novels are larger and more inclusive time capsules than most. ("Heart")

This is a particularly striking compliment when one realizes that it comes from the first author chosen by the Future Library project to compose an entire manuscript, *Scribbler Moon*, that literally will wait a century before greeting its first readers.[1]

Atwood's reasoning does not just concern the aesthetic pleasures of Powers's work, though. Regardless of whether one is a student of the novel, Powers's uniquely informed, evenhanded, and resolute responses to the rapid acceleration of genetic technologies are urgently needed. Let me repurpose some of his own essayistic thoughts—composed before the novels on which I focus—to suggest why. In 2008, he wrote,

> For a very long time, we have been moving from scripted characters to the co-authors of our own lives. The personal genome is one more tentative step from fate to agency, from fatalism to risk management. We are determined not to be determined. The code is loose and always has been. For good or ill, there's never been a bottle that can hold this genie. Yet the dream of molecular management notwithstanding, we are unthinkably far away from ever being able to control the story. The impenetrable texts will have their way with us, in the end. What do we do in the meantime, here, today? Get literate. Read wider. Read deeper. Read more variously, more critically, more suspiciously, more vicariously. Read in anticipation of retrospection. Page one is already being changed by all the pages still to come. ("Book of Me")

This argument suggests that in learning to read its own and other species' genomes, and especially in beginning to rewrite sections, humanity must become better students of its larger stories, from the moon-shot triumphs to the worst atrocities.

Marveling at natural art on every scale, from the nucleotide to the whooping crane, Powers can help. His work never treats species differences in an essentialist, hierarchical, or anthropocentric manner. Where Atwood's *MaddAddam* trilogy is a tragicomic mutation of genetic fantasy, Powers's novels show how genetic realism can grow into its own variety of metafiction by resisting the temptation to pretend perfect mimesis. Earlier chapters explored ingenious metanarratival gestures within realist novels like *Middlesex* and television shows like *Orphan Black*, but Powers joins Atwood in imagining whole works of art as reflections on the potential of genetic fiction. His novels show again and again that humans are transient creatures, not the gods we might delude ourselves into trying to become, yet this is never an invitation to complacency or a rejection of scientific ambition. It is a call to immersion in the moment, to awe at the mystery inscribed in the matter that surrounds us, the matter that we are. Is a bold feminism needed? Badly. Environmental activism? Absolutely. Gene therapy for devastating diseases, healthy debates about additional forms of gene editing? No doubt. But while demonstrating all this, Powers's fiction stretches toward another, even higher frequency. It looks the genome age in the eye and says, "Move forward, but face backward. Don't miss the now."

In that spirit, this section sets the stage for *Generosity* and *Orfeo* via brief looks at two of Powers's earlier novels and his essays about genetic testing. First, consider one of the more trenchant themes in his magisterial 1991 novel *The Gold Bug Variations*, which ranges from the double helix's discovery in the 1950s to the beginnings of the Human Genome Project. Here Powers celebrates the wonder that should motivate scientific work, and I will risk the appearance of redundancy to drive home just how persistently Powers has sought to cultivate this curiosity and resist its diversion toward promises of wealth and power. Relatively early in this lengthy work, the narrator corrects a common mistake he finds among journalists: "Science is not about control. That is technology, another urge altogether. . . . The purpose of science, if one must, was the purpose of being alive: not efficiency, or mastery, but the revival of appropriate surprise" (129). A few hundred pages later, in a friendly argument between scientists, the wisest insists, "Science is not about control. It is about cultivating a perpetual condition of wonder in the face of something that forever grows one step richer and subtler than our latest theory about it. It is about reverence, not mastery" (411). Finally, still another two hundred pages into the tome, the narrator calls science "a way of looking, reverencing" and then reiterates the novel's implicit thesis: "The purpose of all science, like living, which amounts to the same thing, was not the accumulation of Gnostic power, fixing of formulas for the names of God, stockpiling brutal efficiency, accomplishing the sadistic myth of progress. The purpose of science was to revive and cultivate a perpetual state of wonder. For nothing deserved wonder so much as our capacity to feel it" (611).[2] This orientation to the work of the laboratory, this "perpetual state of wonder," has been at the core of Powers's imagination throughout his career, and it is pivotal for understanding what is at stake for Thassa Amzwar in *Generosity* and Peter Els in *Orfeo*. It is the attitude with which both approach the world, in Thassa's case via the frames of video technology and in Els's through his microscopic musical compositions.

The Echo Maker also provides a useful lens for Powers's subsequent novels about genetics. My guess is that this is Powers's most widely read work, given its awards, its relative brevity, its mystery-driven plot, and its moderate level of structural complexity. Set in Kearney, Nebraska, sandhill crane capital of the world, it is a tale of the shifting boundaries between human desires and wildlife needs and between cognitive processes and concepts of reality. The self and its others, it turns out, are necessary illusions upheld by the interaction of specific brain regions, and when their communication is rerouted, the most outlandish beliefs outduel common sense. Mark Schluter's sister, Karin, learns this lesson painfully when his truck ends up in the ditch and he decides several days later that she is an imposter. Because his amygdala and inferotemporal cortex have been disconnected, he knows that she matches the profile of his sister almost perfectly, but he cannot *feel* her as familiar and thus rejects her. Enter Gerald Weber, an East Coast neurologist modeled in part on writer-physician Oliver Sacks, a man who struggles with the requirements of being "Famous Gerald" even as his wife attempts to keep him grounded. Weber finds in Mark a rare instance of Capgras syndrome that lays open "the naked brain" (123), a field study confirming many of his theories about the construction of the self. Unfortunately, though, his fascination leads to objectification. While readers have every opportunity to join in Weber's curiosity about the strange affliction—Mark believes even his own dog has been replaced by some kind of substitute—they also must see how medical science would use the man. Karin senses her brother's disappearance behind the case study, and while grateful for Weber's assistance, her disappointment sends her back into relationships with two men who embody the extremes of preserving nature at any cost and exploiting it for sheer profit. First, she shacks up with vegan Daniel Riegel, the very picture of humility, abstinence, and devotion to his work at the crane refuge; then she goes behind his back to seduce Robert Karsh, a developer who pretends to be interested in raising the town's touristic appeal with a new bird-sensitive viewing center but whose real plans would devastate the ecological balance on which the cranes and their human neighbors depend.

The Echo Maker's configuration of science, environmentalism, and capitalism provides invaluable perspective for understanding the roles played by genetics in *Generosity* and *Orfeo*. The men pursued by Karin typify the extremes of preserving and exploiting nature, but remarkably, the novel ultimately privileges neither. At first, the novel sympathizes more openly with Daniel, whose introverted but openhanded personality partially mirrors that of Powers, but eventually we also recognize the character's dogmatism and naïveté. By simultaneously highlighting Karsh's savvy business instincts but condemning his shortsighted land-use strategies, Powers charts a path toward the same sort of moderate conservationist ethic that new genetic technologies require. Zoomorphizing humans rather than anthropomorphizing animals, the novel anticipates Powers's efforts to imagine biotechnological gifts that embrace but do not essentialize genetic inheritance. Humans *are* animals, we *are* bodies, the soul comes from nowhere if not genetic patterns and neurological impulses—and yet its significance still somehow exceeds measurable statistics. Consider the parallels between *Editing the Soul*'s overarching questions about genetics and Weber's about cognition: "How does the brain erect a mind, and how does the mind erect everything else? Do we have free will? What is the self, and where are the neurological correlates of consciousness? Questions that had been embarrassingly speculative since the beginnings of awareness were now on the verge of empirical answer" (134). This is the same onrush of biotechnological possibility that motivates Powers in *Generosity* and *Orfeo*. There the questions merely focus on how genes become proteins and how these become persons capable of asking questions about their own origins.

Two further elements of *The Echo Maker* also deserve attention before moving ahead. The first is neatly summarized by the novel's epigraph, a quote from Soviet neuropsychologist A. R. Luria: "To find the soul it is necessary to lose it." This might seem too tidy an existentialist paradox, but Powers returns to it again and again. At the obvious level, Luria's words suggest how Mark has to lose parts of his memory and his ability to recognize loved ones in order to appreciate them more later. But the characters who journey furthest into hell and back in this novel are the ones who *observe* the neurological testing. Karin must recognize how much she has allowed others' approval to determine her most crucial decisions; Mark's doctor, Gerald, must grasp that he has become more interested in his patients as illustrations than as fellow beings and that he is drawn to what could only be a brief romantic fling, even if it destroys the decades-old narrative he has been composing with his wife. In short, my postsecular reading of Powers's emphasis on remaking the soul is that Karin and Gerald must consciously decline short-term temptations in order to renew the deeper threads of their lives. This does not make the novel "religious" or "sentimental" in any traditional sense, but unlike Dawkins and other militant atheists, Powers does not wish away all forms of spiritual transcendence—no more so than he suggests reducing a human being to a genetic bar code, even as he agrees this information can sometimes prove enormously valuable.

A related, equally counterintuitive maxim of *The Echo Maker* is that "logic depends upon feeling" (106). For those trained in the sciences, this can seem backward: one should establish the facts first, only then (and separately) examining the feelings they create. But Powers is not appealing to sentimental whims or peddling a cheap, stereotypical relativism. Rather than approaching reality as a smorgasbord to pick through as one desires, his idea is far more radical and foundational to the history of science and epistemology, particularly as these fields have been revised by feminist scholars in recent decades.[3] As detailed in chapter 1, objectivity is always limited, humility is always necessary, and the observer always influences the observed. Subjectivity demands awareness of one's limitations, not pretended certainty. Of course, rigorous laboratory work requires circumventing every possible source of variability and ambiguity; fuzzy data and low replicability are among the fastest routes to experimental insignificance. Yet what Powers urges is the realization that even as a scientist proposes an experiment or begins to interpret the data—not to mention as committees debate whether to fund it or journals decide whether to publish it—the processes at work are far from cut-and-dried or uninflected by human desire. The culture of science is strongest when its human motives, assumptions, and even uncertainties are directly acknowledged, not glossed over. This is what Gerald Weber rediscovers as he returns to Nebraska on research trips that have as much to do with Mark's nurse as the Capgras patient himself. Only on recognizing his mixed motives can he confront his own neurological short-circuiting and self-deception.

Weber's dawning awareness of his infatuation sheds more light on the plot of Powers's next novel than might be obvious. The remarkable aspect of Mark's nurse is that she exudes such tranquility and such endless patience with the physical and emotional needs of a man who is far from the model patient. Karin reflects that she has "the lightest touch known to man" (241). As we eventually learn, though, Barbara's indefatigability has an explanation. Not only was she the near suicide who stepped in front of her future patient's truck; she was in the region on false pretenses, hiding her identity as a television journalist in an effort to investigate the very water access problems that Karsh's development would exacerbate. In *Generosity*, Powers effectively splits Barbara into two characters so as to pursue these narrative threads even further. Tonia Schiff wrestles with the same challenges as Barbara in separating her storytelling integrity from her popular science show's emphasis on short-term profits and ratings, while Thassa Amzwar is an even more impossibly magnetic personality than Mark's nurse, one whose exuberance is so attractive that it seems proof of a "happiness gene." In both cases, as with Barbara, Powers eventually exposes the illusion. Tonia can only remain innocent for so long before she must make a choice about how far she will let herself be used. Her colleagues are willing to edit an interview to say anything, even the opposite of what the raw footage clearly indicated, but inspired by the spectral creature in the woods, she finds the mettle to follow her conscience, walk away from her show, and start over. Ultimately, this means following Thassa back to her home country, Algeria, in order to hear her side of the story, to learn how the woman everyone envied could have destroyed that perfect image so violently. Indeed the former art student reaches her own breaking point: when everyone from the press to a megachurch pastor interprets her genome test as evidence of biological fate, when they try to make a messiah of her, she flees. In Powers's narrative experiment, if you make the happiest woman alive famous for her genes, the joy wears thin. Even if the tests are accurate and Thassa is as much an outlier as she seems, genes can only decide so much.

The Echo Maker also foreshadows *Orfeo*'s treatment of do-it-yourself genetic engineering. If *Generosity* is a fictional investigation of personal genome testing—what knowing one's microbiological inheritance can and cannot reveal—*Orfeo* is its corollary for a growing biohacking movement. While the geneticist in the 2009 novel is modeled in some part on George Church, Powers's 2014 novel offers a musical riff on the experience of SUNY Buffalo art professor Steve Kurtz, who was arrested by the FBI on suspicion of bioterrorism (but not charged) after he called 911 to report his wife's death and police officers saw various bioart materials in his home. In *Orfeo*, Peter Els reports the death of his dog, but he is away when the FBI raids his home and flees when he comes on the scene. His subsequent journey might be understood as an extension of the automotive flight from reality that Gerald and Barbara begin in *The Echo Maker*, except that Els's flight is toward reality, one to which his nation's fear-driven bureaucracy seems deaf. The same can be said of Russell and Thassa's car trip near the end of *Generosity* and her eventual return to Algeria: the protagonists of *Generosity* and *Orfeo* have seen things that society as a whole is unready to face, yet they are running toward truth, not away from it. Whereas Gerald and Barbara must turn back in order to help Mark and expose the duplicity of a supposedly environmentally friendly tourist attraction, Powers's characters in the more recent novels must escape rabid fans and seething enemies alike. Only from the other side of national borders or behind the walls of social media can they uncover the interpretive errors that portray the genome as fate, that assume microbiological curiosity makes one a terrorist. These novels demonstrate that for all the new millennium's advances, Western culture remains woefully unprepared to grasp the significance of new genetic tools and the microbiological frontiers they open.

This lack of preparedness extends to many artists, but for a growing minority, it constitutes a call to action. When Powers called for greater genetic literacy in his 2008 essay "The Book of Me," he showed again to which group he belongs. As is also evident in the 2006 essay in which he anticipated Barnes and Dupré's "astrological" metaphor by describing personal genomic testing not as "palmistry" but at best "informed weather prediction," Powers closely monitored the early developments in this industry. Well before his own genome test, Powers sought to shift readers' expectations from "divination" toward a more proactive mind-set, recognizing that even on the rare occasions when tests point to definitive single-gene disorders, "knowing what's coming does not shield us from living it." His interest was less the fates these screenings implied than affected individuals' responses: "Tests that increase a patient's ability to write her own life are deeply desirable," he reasoned, but "tests that decrease the patient's ability to write her own life are not. Which tests do which depends upon their taker" ("A Brief Take" 46). Before becoming one of the first eight individuals in the world to have his genome fully sequenced, that is, he already knew the data's significance would require considerable interpretive labors. At that point, commercially available tests sampled only tiny portions of customer genomes, whereas in exchange for Powers's essay, *GQ* paid six figures for twenty-three scientists to test close to 100 percent, completing ten separate sequencing runs in order to increase accuracy. Powers enjoyed a unique vantage point as he began writing *Generosity*, and it made him highly conscious of the distance between information and meaning.

Not surprisingly, "The Book of Me" reads something like an abstract for Powers's next novel, with the author not only wondering "if we aren't in danger of pathologizing ordinary health, turning us all into pre-patients for diseases we are only at risk of contracting" but also acknowledging the potential benefits of such tests. The essay anticipates a growing discussion about "previvors," a term that has emerged for those who respond aggressively to genome test findings with preventative surgeries and other interventions. Powers's main interest, though, was in psychological effects due to "the gradual replacement of luck with control. Once upon a time, we were dealt a hand by Fate, God, or the Unreliable Narrator, and the task of life was to deal with that hand. Now the task is to improve the deal." Characteristically explicating new biotechnological developments via literary tropes, he summarized, "For a very long time, we have been moving from scripted characters to the co-authors of our own lives. The personal genome is one more tentative step from fate to agency, from fatalism to risk management. We are determined not to be determined." In other words, after enduring the suspense of learning about his individual proclivities to certain diseases, what interested Powers most about the process was its implications for human control of biology—or rather, the lack thereof. He recognized that the process was growing more accurate and incisive and that there would eventually be real ways in which it could serve patient health, but he kept a constant eye on the temptation to surrender to illusions of overwhelming medical force. To judge by his essay, the net result of his experience was a greater awareness of how the more control one achieves, the more seductive its overstatement becomes.

The rest of this chapter examines how *Generosity* and *Orfeo* teach us to take advantage of biotech blessings while resisting one-dimensional treatments of their impacts. The first novel follows a young immigrant who infuses others with life but only barely survives their endless need, their readiness to make her a messiah and then a scapegoat.[4] The second is a tale of an older man who wonders if he was wrong to abandon chemistry for music in his youth. Now he only wants a few more glimpses of the genetic splendor that his secondhand equipment can divulge, perhaps even the chance to write something that will outlast the avant-garde music he has committed to paper. Neither character escapes their chaos in triumphant fashion, but both Thassa's and Els's stories are landmarks in the evolution of genetic fiction. Oscillating between these works, the remaining sections pursue two of the largest questions in *Editing the Soul*. First, according to these novels, how might humanity best use its rapidly growing knowledge of the genome? What does it mean to incorporate such information wisely, and how can we do so without selling our souls, or put more soberly, sacrificing our ethics? Second, what do humanity's near-term biotechnological capabilities have to do with the purposes of fiction? According to these stories, what constitutes free will as knowledge of biological inheritance accelerates, as modifying it grows ever more plausible? Whether querying the purposes of science or literature, Powers offers no quick or easy answers, but he makes clear that within both novels and cells, the challenge is to rewrite lives in ways that remain open to revision and eschew performed certainty, ways that reach beyond the control delusion cultivated by technotranscendence, ways that more fully embrace the wonder of being alive at all.

From Selling Genetics to Modest Witnessing

This book began with George Church's talent for taking science public. Whether on a late-night talk show or a panel at a Modern Language Association conference, he is an engaging, witty conversation partner and a very effective ambassador for contemporary genetic innovations. Thomas Kurton, the prominent geneticist in *Generosity*, is not George Church, just as the novel's Chicago is not the actual city, but as the narrator puts it, "Chicago's in vitro daughter, genetically modified for more flexibility" (6). Kurton is not Church any more than Gerald Weber in *The Echo Maker* is Oliver Sacks; rather, the actual individuals only provide templates for fictional entities that Powers allows to mutate considerably across their narratives. Still, in making sense of Powers's fictional geneticist, it is well worth recalling the virtuosity of Church's interview with Colbert. It was Church's lab to which Powers traveled in order to begin his own genome sequencing process, so Kurton might be understood to reflect what someone in Church's position *could* become. For all his faults, that character remains likable, a Martin Arrowsmith for the twenty-first century. Admittedly, he can be intolerably narcissistic, from the cryogenics bracelet to the dietary extremes to the appraising glance at Tonia's hair; still, this is no mad scientist but a man who was once a child and has not entirely forgotten the experience. I begin this section with Kurton because he epitomizes both the enormous possibilities that attract Powers to genetics and what he cautions readers to avoid becoming. He is a person of enormous vision and imagination, yet one deluded about his ability to determine reality and oblivious to the limitations of his empathy. The next section shifts to Thassa as well as Peter Els in *Orfeo*, characters with far greater humility but no less curiosity. Through this juxtaposition, my aim is to show how *Generosity* and *Orfeo* reveal the capacities of genetic biology both to do great good and to cause great harm, potentially at the same moment.

It would be easy to misunderstand Kurton as the straw man against whom I am praising Thassa and Els, so I want to speak first in his defense. Most critically, as Thassa confirms by taking Kurton to see the leafy sea dragons at a Shedd-like aquarium, he retains a capacity for genuine astonishment, the attitude that Powers has suggested for decades is central to true science. He recognizes these organisms as "pure possibility, feeling how feeble imagination is, alongside evolution" (*Generosity* 145). Even as an eighth-grade boy—not always the most likely creature to exude awe in the science classroom—the future geneticist sensed that "nearby species were already more alien than any fiction." Dissecting frogs and finding "life's true measurements" (41) under the microscope, Kurton had already decided that "the love he really lived for was *knowing*" (43, emphasis in original), a mind-set that recalls Arrowsmith and his obsession with finding the "X Principle." That 1925 novel's scientist also "wanted to know—*to know*" (Lewis 322, emphasis in original), and in Kurton's mind, nothing is better: "To look on a thing that had been true since the start of creation but never grasped until *you* made it so: no euphoria available to the human brain could match it" (Powers, *Generosity* 43, emphasis in original). In short, Kurton's intellectual curiosity is authentic, even if it comes with an unusually insistent ego. He is no dead empiricist, grasping how crucial it is to follow his gut; "hunch's role in science has never embarrassed Kurton" (128). Rather than fulfilling definitions of science that ignore intuition and educated guesses, the geneticist risks losing his research project's primacy in the hopes that a case study like Thassa might come along to flesh out the numbers more convincingly.

Before judging Kurton too harshly, it is also worth remembering that he faces an uphill battle with a large segment of American culture. For scientists who depend on public research dollars, there is a fine line between confronting ignorance and remaining diplomatic. Kurton is understandably disgusted by the fact that the basics of natural selection can be found "anywhere, aside from a quarter of American high schools" (146). As he laments on one of Tonia's TV show episodes, "We live in a country where 68 percent of folks don't believe in evolution" (20). In our world, this is a moderate overstatement, but whatever the current statistic, he is right that it is very difficult to help citizens grasp the significance of the human genome when they reject out of hand the paradigm by which humanity has started to understand inheritance in the last two centuries.[5] With that in mind, Kurton does an effective job of balancing salesmanship and precision as he discusses genetic applications. He does not announce baldly, "We've found the happiness gene" (even if he knows this is how his words will be interpreted), but only that "this network of genes seems to account for perhaps two-thirds or more of the heritability of emotional temperament" (122). The rules of the funding game demand couching enormous implications in the most careful language possible, and Kurton is an expert at keeping his balance. As *Generosity*'s narrator puts it, scientists like him "had simply accepted science's latest survival adaptation—salesmanship," and so "any funded researcher who condemned them was a hypocrite" (135). The implication is that if US society wants straighter talk from its laboratories, it needs to be much more ready to reach beyond glib talk-show rhetoric and, as Powers put it in his essay, become literate in the biological fundamentals.

Nonetheless, Powers also offers a penetrating critique of his character's bad faith, even when it seems unconscious. Kurton might not be "the thuggish Edward Teller nor the grandiose Craig Venter that scared or envious reporters made him out to be," but he is still "vaguely messianic" and inspires thousands to preach his "clergyman-chemist's ecstatic vision" (135). As earlier chapters have shown, the transcendent language he employs is endemic to many forms of posthumanism. Kurton is far more sophisticated than *White Teeth*'s Marcus Chalfen or *Orphan Black*'s Aldous Leekie, but he still signs e-mails with the triumphalist tagline, "Whatever the beginning of this world, the end will be glorious and paradisiacal" (136). Even Thassa, the gentlest of critics, recognizes that "he knows how to make the donkey think it's choosing the rope" (142). Kurton proves especially captivating in his filmed interview with Tonia, losing himself in a disquisition about the powers of personal genome testing. Seizing her coffee cup and taking a quick swab of its backwash, he proclaims, "I now have your genetic profile. Your SNPs and indels—the variations in your genome of any significance. I can identify your ancestors—and your descendants. I can predict your health and development, and I can even speculate about your disposition. I can make a good bet of your likely age span and what you'll die of, if you don't get hit by a car first" (173). This makes for solid entertainment, both because Kurton is so confident, and perhaps also because of the sexual undertones of his theatricality—after all, he has just taken something from an attractive female interviewer on screen, without permission, knowledge that could hardly be more intimate. Yet he speaks so urgently, so charmingly, that it seems possible to overlook the violation, at least temporarily.

Temporal context is a major factor here. If Jay Clayton has demonstrated how genome time registers the changes in perception that follow from humanity's accelerating acquisition of microbiological knowledge, Powers's novels provide invaluable illustrations that fold in the effects of contemporary news cycles and edutainment expectations. A quintessential scene is the *Oona* talk show taping in *Generosity*, in which Thassa meets Kurton for the first time since news of her unique genome goes viral. Powers lets readers see what is happening through the frame of the TV camera viewfinder as well as from Russell Stone's seat in the audience. A lightly disguised adaptation of the long-running *Oprah* show, *Oona* pretends to enter an everywoman living room in which any viewpoint can have its day, so long as it is ready to face public cross-examination. Powers portrays this soundstage of contemporary mythmaking as the pseudointellectual environment in which American culture will make its most crucial judgments about emerging biotechnologies. There is a clear agenda here: Oona's "belief that fortune lies not in our stars but in our changing selves" (210) and her reaction to a biologist pushing "a predisposition to disposition: it's exactly the kind of fatalism the boss is determined not to be determined by" (211). There is also considerable stagecraft: before the host or guests appear, audience members are urged to "be themselves and respond honestly to any meltdowns that Oona and her guests get into," but in the very next breath, they are informed that "monitors spread throughout the room will give simple cues to help indicate where laughter or surprise might be appropriate" (215). Then the audience practices their collective responses to the screen prompts, becoming a homogenous, malleable organism ready to follow producers' cues.

Generosity thereby emphasizes that few people know quite what to feel about new forms of genome testing and gene editing, and as result, they are very amenable to suggestion. On one hand, relatively few citizens will be captivated by the latest James Bond scientist villain seeking world dominion, nor by the stereotypical Luddite fundamentalist who would send the world back to the Dark Ages. Beyond rejecting the absolute extremes, however, there is considerable befuddlement, and it has left an enormous opening. In Powers's storyworlds, shows like *Oona* rush to fill these vacuums in public opinion, nudging citizens to decide whom they trust, not necessarily to evaluate arguments with the greatest care. Psychologist Candace Weld is therefore a critical conversation partner for Russell as he struggles to interpret Kurton's findings for himself. Even before the *Oona* spectacle, Candace helps Russell see that it really doesn't matter how good the biology is in Kurton's paper about Thassa and the genetics of happiness, because "every conclusion in the article could be discredited next month, and journalists will still be reporting it five years from now" (166).

In other words, the media spotlight makes the journal article's *interpretive work* invisible. Once sliced into brief quotations by media outlets, there is no uncertainty left, just the comfortable appearance of lab-coat authority. Even for someone who knows the subject well, "before Stone's eyes his sunny former student turns into a genetic aberration, immune to disaster, a product of chemical reactions qualitatively different from his" (168). The pseudoscientific public discourse alienates Thassa in ways even her ethnicity and war-torn childhood could not. Kurton might have been careful with his wording, but his complicity with journalists' hunger for spectacle contributes to making "two-thirds of the [website-]commenting public believe that nature contributes more to happiness than nurture, up from 50 percent a year ago" (183). This is the same shift toward overemphasizing biological inheritance that we heard Jeffrey Eugenides lament in chapter 3. As Russell realizes while watching the *Oona* taping, once a scientist like Kurton has such a platform, "the man can say anything at all. Sober measurement or wild prediction—it makes no difference. He's on the show. And the show, not the lab, is where the race will engineer its future" (217). Indeed it is not a scientist's facts that matter so much as looking like an expert. Shedding two decades from his physique with a careful diet, masterfully combining evasiveness and levity, Kurton is exceedingly well adapted to this environment. Ironically, it allows him to overemphasize the impact of genetic inheritance, illustrating again how so-called genetic determinism is really a culturally promulgated ideology and more a reflection of nurture than nature.

"The show" also plays a major role in *Orfeo*, where Powers invites readers to identify with a less image-savvy hobby scientist. In this case, the hapless Russell Stone–like character becomes the scientist and the focal technology is not personal genome testing but garage-based biohacking, the DIY movement by which anyone with patience, a thousand dollars or so, and the ability to follow online recipes can pry open a world of DNA previously accessible only to specialists. Unfortunately, this possibility is news to the cops who show up at Peter Els's home. In an America beset by threats of terrorism, when Els flees the ensuing FBI investigation, the media makes him into "the Biohacker Bach," leaping to the conclusion that his experimental strain is responsible for a spread of bacterial infections in Alabama hospitals (265). The real cause turns out to be a contaminated batch of IV bags, but as with Kurton's performance on *Oona*, the facts are lost in the spectacle. The fear even infects Els, so that every security camera, every credit card slot, every cell phone consultation represents the threat of capture, when all the retired music professor wants is a chance at an organic alternative to the twelve pitches of the chromatic scale. Els's reasons are those of the artist, and as such they are completely unfathomable to the news anchors and bloggers out for blood. Little about bioart is credible to many nonparticipants, but "in the year of Els's birth, no one had even known what a gene was made of. Now people were designing them. For most of his life, Els had ignored the greatest achievement of his age, the art form of the free-for-all future that he wouldn't live to see. Now he wanted a little glimpse" (46). For that curiosity, and for his creativity, the musician pays dearly.

Whereas Powers demonstrates in *Generosity* how easily an enormously complex, polygenic, environmentally affected trait like demeanor can be reduced in the court of popular opinion to the inheritance of a single gene, he shows in *Orfeo* how easily a gentle, curious man pursuing a project with no commercial interests or health implications whatsoever can have his character assassinated by media-stoked fears of bioterrorism. In both cases, Powers pushes back against the misperceptions by reemphasizing humanity's creaturely status. Tonia's arrest by the unnamed animal beyond Kurton's house is repeated in *Orfeo* several times over, via species both visible and invisible. Els's most profound testimony in the novel, in fact, is that "the environment" is not "out there" but "in here." This is evident from the novel's first pages with the death of his musical dog, Fidelio, which leads to an illegal burial in his backyard, the dog wrapped in a quilt made by Els's ex-wife. A scene that could have been sentimental is instead representative of the empathetic dog-human bonds in Peter Heller's novel *The Dog Stars* (2012) or Donna Haraway's *When Species Meet* (2007), bearing out Els's reflection that "if humans had a soul, surely this creature did. And if humans did not, then no gesture here was too fine or ridiculous" (12). The theme extends throughout the novel, as when Els recalls a walk with Fidelio in which he was dazed by the "rhythms" in a wet oak leaf and looked up to sense how "music floated across the sky in cloud banks, and songs skittered in twigs down the staggered shingles of a nearby roof. All around him, a massive, secret chorus written in extended alternate notation lay ripe for transcribing" (331). Powers refuses naïve idealization, but his plant and animal worlds are just as fully alive as the human one with which they interweave.[6]

Powers also brings this awareness of natural glory to *Orfeo*'s passages about life on a microscopic scale. Els is both creator and audience in his homemade lab, where he is stunned into silence by DNA replication: "For two hours, DNA melts and anneals, snatches up free-floating nucleotides, and doubles each time through the loop. Twenty-five doublings turn a few hundred strands into more copies than there are people on Earth.... The man puts his eye up to the lens and sees the real world" (2, emphasis in original). The portal to this deeper reality never closes for Els. DNA's doubling and redoubling "made him feel religious" (42); "the formulas of physical chemistry struck him as intricate and divine compositions" (57); he finds, "in a single cell, astonishing synchronized sequences, plays of notes that made the Mass in B Minor sound like a jump-rope jingle" (142). Even in his last conversation with his ex-wife, he realizes that words remain insufficient: "He couldn't begin to tell her. Life. Four billion years of chance had written a score of inconceivable intricacy into every living cell. And every cell was a variation on that same first theme, splitting and copying itself without end through the world. All those sequences, gigabits long, were just waiting to be auditioned, transcribed, arranged, tinkered with, added to by the same brains that those scores assembled. A person could work in such a medium—wild forms and fresh sonorities. Tunes for forever, for no one" (299). If nothing else is clear about Powers, it should be that his protagonists embody the best traits of Haraway's "modest witness." Far from a hyperabstracted, heroic knower, Els is a humble participant in the world he finds beneath the microscope, and in that sense becomes a model for readers, despite his occasional awkwardness with other people and his youthful stupidity in abandoning his family for short-term fame.

This guilelessness is what distinguishes Els most fully from Kurton in *Generosity*. Kurton can gaze upon the leafy sea dragons and be reminded of endless forms most beautiful, but his attention swiftly returns to talking a would-be research subject into her specimen container. For Kurton, the joy of knowledge is displaced by his achievement in acquiring it—ideally, before anyone else. This comes through all too bitterly in his final interview with Tonia Schiff, as he rests a case for gene patenting on the claim that nothing could be more valuable than knowledge: "A few slavish chemicals producing damn near omnipotent brains.... That discovery is better than any drug, any luxury commodity, or any religion. Science should be enough to make any person endlessly well. Why do we need happiness when we can have knowing?" Yet this is precisely where his case breaks down: Kurton lets slip that ultimately, he is uninterested in joy or the specific woman, Thassa, who so fully embodies it. Instead he assumes that life has only gained purpose now that he is around to measure it: "Six hundred generations ago, we were scratching on the walls of caves. Now we're sequencing genomes. Three billion years of accident is about to become something truly meaningful" (252). This condescension toward earlier eras is almost as sad as Kurton's unawareness that his work has made Thassa the exact opposite of "endlessly well." As the narrator summarizes in Kurton's last scene, "All life long, he has believed in the one nonarbitrary enterprise, fairer than any politics, truer than any religion, deeper than any artwork: measurement" (271). The grim reality is that Kurton has swallowed whole the core of scientism, an ideology that too often steals attention from the very methodology it aims to honor.

By contrast, one of the last scenes in *Orfeo* dramatizes Els's opposite attitude toward science's purposes. Reunited with a long-estranged friend, Els is invited to peer through a telescope and appreciate a far larger set of constellations than he normally studies. The effect, though, is familiar: "Els puts his eye to a burst of stars. They cluster, a blue star nursery, spraying out new worlds. He feels like he did two years ago, when he first looked at a glowing stain of cells under the 1,000× objective and realized that life happens elsewhere, on scales that have nothing to do with him" (347). The shift in scope is one more reminder of Powers's insistence that life matters deeply, in and of itself—that contra the ideology of this world's Kurtons, there is already unfathomable meaning in the cosmos. Also, unlike more vapid proclamations of humanism, Powers does not make value dependent on human centrality. The emphasis on "scales that have nothing to do with [Els]" echoes the many scenes in which Powers takes this conviction to the opposite extreme, not countless light-years away from the earth, but down to the scale of the nanometer. His work is a response to the need he expressed in a conversation with Scott Hermanson and Tom LeClair in 2008: "If we could wrap our heads around the fact that life has evolved from inanimate polypeptides to something that's put a probe on Titan that's sending back pictures, we might be more motivated to summon up the collective will to stave off our own extinction. But for that, we need to be taken out of our private, unconnected lives. We need stories that can change the scale of our identification" ("Home").

In a similar vein, Stacy Alaimo criticizes an environmentalist mind-set that values other entities according to their distance from the human, and she offers a compelling alternative: "Trans-corporeality denies the human subject the sovereign, central position. Instead, ethical considerations and practices must emerge from a more uncomfortable and perplexing place where the 'human' is always already part of an active, often unpredictable, material world" (16–17). In celebrating Els's discovery of his dependency on a microbial world to which he was once blind, Powers effectively aligns his novel with Alaimo's transcorporeal alternative:

> A year of reading, and the scales fell from Els's eyes. Bacteria decided wars, spurred development, and killed off empires. They determined who ate and who starved, who

got rich and who sank into disease-ridden squalor. The mouth of any ten-year-old housed twice as many bugs as there were people on the planet. Every human body depended on ten times more bacterial cells than human cells, and one hundred times more bacterial genes than human ones. Microbes orchestrated the expression of human DNA and regulated human metabolism. They *were* the ecosystem that we just lived in. We might go dancing, but they called the tune. (193, emphasis in original)

By engaging genes beside star clusters—not just using the science as a metaphor for some aspect of identity, as effective as that can be—Powers acknowledges humanity's midrange perspective but refuses to deify it. Paying close attention to bacteria is not merely an environmentalist conceit but a necessary step toward humility, toward recognizing ourselves as stardust. That it transforms one's vision is evident in many moments in *Orfeo*, but none is more powerful than Els's description of another human being as a "perfect, working creature, self-assembling, self-delighting, the brightest whim that could ever exist" (170). This is his portrait not of a bacterium but of an organism he loves especially dearly, his daughter. Such an oblique description is one more instance of Powers's modest witnessing, one more expression of an enduring sense of wonder. Analyzing *The Echo Maker*, Heather Houser insists that such openness to awe allows humanity to "become conscious of perception itself," to "marvel not only at the astonishing object but at our own capacity for awareness and our position relative to larger systems" (388). The next section reveals how the very form of Powers's novels enhances that potential.

Predisposed Agency and the Joy of Not Deciding

Powers's 2008 essay about his genome testing experience includes this reflection on literary structure: "In *Reading for the Plot*, Peter Brooks suggests that 'our chief tool in making sense of narrative, the master trope of its strange logic,' is 'the anticipation of retrospection.' Page one means what it means only because we already know that page 300 is going to change it forever." This gloss on Brooks offers a pathway to understanding the self-reflexive structures of Powers's novels. In *Orfeo*, for instance, "each measure [Els] wrote changed the ones he'd already written, and he felt them all, already being altered by the unformed noises of the years to come" (62). There is no simple linearity here, no easy "begin here," "end there" instructions, but a consciousness that the now is constantly re-created by a formerly present moment being swept by the tide into history, by the formerly future wave rushing into immediacy, thereby creating a new set of possible futures. To recall Brooks once more, "an event gains meaning by its repetition, which is both the recall of an earlier moment and a variation of it: the concept of repetition hovers ambiguously between the idea of reproduction and that of change, forward and backward movement" (99–100). Rather than simple linear progression or circular regression, Brooks and Powers are describing an undulating, spiraling sense of time, an imbrication of sequence and simultaneity. In this section, I spell out why that synthesis is so significant for understanding both genetics and fiction and how it can make interpreting a genome and interpreting a work of art strangely parallel tasks. Taken together, *Generosity* and *Orfeo* are literary testimonies to the agency built into even the most specific bodily and narratival predispositions. They are invitations to live by myth, not over against science, but in recognition of these methods' interdependence—to live, write, and study life with an eye to discovery more than invention; to welcome instead of fearing uncertainty; and to do so precisely by taking risks from within it.

Immediately following the narrator's observation that "each measure he wrote changed the ones he'd already written," *Orfeo*'s study of its protagonist continues: "As his pencil spilled lines onto the blank page, all Els had to do was listen and guide each new note to its foreordained place: *They are alive and well somewhere*. He could write for forever; he could write for no one. He wasn't choosing: simply detecting, as if he were running a dozen different assays to determine an unknown that, by reactive magic, precipitated out in the bottom of this test tube, solid and weighable" (62, emphasis in original). Powers is full of unusual comparisons: uncovering the next musical note in the pattern, coaxing a chemical solid from behind its liquid mask, watching to see what a character will say and do next. Each task requires parsing the true and false notes, unveiling the pattern beneath apparent chaos, whether hidden in a test tube or behind a blank page. This is a unified theory of making art and doing science, a loose methodology that relies on the conviction that in some way, humans know things before we know them, that hypotheses in the lab, musical innovations, and intuition in forming characters are imbricated at a subconscious level. Powers draws together the most creative of scientific theorizing and the most rigorous of literary and musical creative work as the same sorts of mystical labor. These processes are far easier to marvel at from afar than to undertake oneself; they are much easier to undertake briefly than to sustain.

Generosity and *Orfeo* are largely about characters who ride their waves of intuition far longer than most. Thassa is the fictional incarnation of Powers's 2006 essayistic conviction that "genetic tests are not about escaping the story; they are about figuring how best to be in it" (30). She is dubbed "Miss Generosity" by her art college classmates because she so deeply evokes the best in each of them. In some measure, she *has* escaped a horrific story: raised in a world of militant Islamism, she saw her parents murdered for their "congenital hope," their refusal to believe that their country would devolve into civil war (29). But she has also figured out, at least more than most, how to live well within her story—that of a young woman who has found a new country in which to make art. Thus her journal entries represent "something out of the dawn of myth, set in a Chicago all but animist" (25), and she appears in Russell's classroom "glowing like a blissed-out mystic" (35). The vitality she exudes is intoxicating, and for a time, she seems impervious to others' hatred and envy, even convincing a would-be rapist that he is hurting *himself*. Eventually, though, the combination of fame from Kurton's study and fears of abandonment by her mentors slams Thassa to earth. All the genetic predisposition to bliss in the world is not enough to protect her from the whiplash of being elevated as an angel and then rejected as a demon.

However, my suggestion is that at its deepest layer, *Generosity* is less about Thassa than Russell, which might seem surprising given that she is the character with the exceptional genome. Yet even as she worries about others' assumptions about her genes, Russell is struggling just as desperately to regain his narrative voice. An adjunct professor teaching "creative nonfiction" in a gig he can scarcely believe he was offered, Russell was once a younger man whose stories began selling faster than he could finish them.[7] There was a whimsicality, a bite to their absurd character sketches that was catnip to editors and readers—at least until he heard from the family of an actual person on whom he had based a character. Blaming himself for damage done by his portrayals, he grew careful, generic, eventually accepting work as a magazine editor that did not spark his imagination but that at least did no harm. In accepting the teaching gig, though, he once again exposes himself to surprise, to students who demand to know whether he lives out what he teaches. As a character, the answer appears to be negative, but what persistent readers eventually must discover is that Russell is the entire novel's camouflaged narrator. Once this realization is made, it changes nearly everything within *Generosity*'s storyworld. To use Powers's terms, page one is dramatically reframed by page three hundred.

This interpretation answers many questions, but it is not obvious. In a 2012 volume of Powers criticism published in Germany, Heike Schaefer considers the same evidence and reaches a very different conclusion. For Schaefer, the novel is an attempt to fulfill Kurton's desire for a "fact-oriented," "post-genomic fiction" that changes radically in order to "remain relevant to the lives of its authors and readers in the age of consumer genomics, commercial TV, and social media" (268). That is, transformations in media and biotechnology "have rendered the conventions of realist, modernist, and also postmodernist fiction obsolete" (269), so "the narrator finds himself in agreement with Kurton who hopes 'to live long enough to witness a new, post-genomic fiction'" (271n3). By contrast, my position is that *Generosity*'s narrator certainly considers the need for fiction to evolve and even finds Kurton's perspective compelling at times, but that ultimately, the novel constitutes a long *rebuttal* to Kurton's assumptions. Crucial to this claim is the realization that while the novel appears to be about characters named Russell, Candace, Thassa, Kurton, and Tonia, on another level it is about the very process by which a character named Russell finds the courage to write a novel about these characters, including the one based on himself. Remember, Powers is the same novelist who wrote *Galatea 2.2* (1995), in which the main character is named "Richard Powers." Clayton summarized the even more complex plot of *The Gold Bug Variations* as follows: "Here are two characters *in* a story writing the story that contains them. The text becomes a strange loop, a recursive process without beginning or end" ("Genome" 48, emphasis in original). My argument might be understood as a reapplication of Clayton's reading to a new novel, one that shifts the narrative shape from a circle to a spiral. *Generosity* features a protagonist writing a story that contains him, creating a work that simultaneously contains a linear narrative and relies on a circular, "creative nonfictional" account of its creation.[8]

From the novel's beginning, Powers offers asides that might seem familiar to readers of postmodern fiction, in which he announces his concern with the purposes of literature while also rebelling against simple genre categories. The narrator comments, for instance, "I never seek out uncanny plots. I find them way too cheaply gratifying. I stay away from books with inexplicable coincidences, prophetic events, or eerie parallels. But they seem to find me anyway. And when I do read them, however conventional, they rip me open and turn me into someone else. This is what the Algerian tells me: live first, decide later. Love the genre that you most suspect" (70). Who exactly is this narrator Thassa advises to plunge into science fiction, horror, fantasy, mystery, romance, superhero comics—all kinds of stories—rather than merely judging them from afar? More clues come when Russell talks with Candace about her preadolescent son Gabe's reading preferences. Following a section in which the narrator ponders how many of the millions of novels printed so far are "saddled with a romance" and what makes people "the fiction-needing readers we are today," Russell recognizes Gabe's fascination with anything "far away, in some parallel universe. A thousand years before or after, anywhere but now." Candace nods, remarking on her son's "infinite hunger for the unreal," and asks, "Why should that be useful, in little boys?" (95). Use-value might not be the right question, but unlike many novelists, Powers is not satisfied with merely posing metanarratival questions. When Russell's role as narrator becomes clear, Powers's entire text becomes a partial answer.[9] *Generosity* is a direct response to its own narrator's fear that "fiction is obsolete" and "engineering has lapped it," a worry regularly apparent in the endless defenses of the humanities and literary studies that characterize our era of increasingly corporatized academia (152).

If the secret that transforms one's reading of the *MaddAddam* trilogy is that Crake is not quite the lone instigator of the Waterless Flood that he seems, the secret of *Generosity* is that the bumbling, fearful adjunct professor who struggles so mightily to retain his students' respect, who nearly proves accessory to one's suicide, is slowly rediscovering the freedom to write his own life. The clues multiply the further one reads, but one of the least dubitable comes when Candace probes Russell, "What would his book be about, if it dared set foot in this world?" He doesn't say immediately, but the free indirect discourse suggests it might be "about the catastrophe of collective wisdom getting what we want, at last"—that is, the very problems that would ensue from being able to modify at will our genes and those of our offspring. On the phone later, she challenges him, "Close your eyes and write a sentence in the air," and he imagines, "*They sit and watch the Atlas go dark.*" When she asks if now he wants to know "what happens next," he moans, "I'm afraid that *was* the next." Rather than accepting this evasion, she suggests, "Then write what happens just before" (152). This conversation reshapes in advance the final sentences of *Generosity*. As the narrator sits across from Thassa on a narrative stage swiftly going dark, he is finally past the need to save her or to use her and can ask simply, "How *are* you? How do you *feel*?" Then come the final two sentences of the novel: "She answers in all kinds of generous ways. And for a little while, before this small, shared joy, too, disappears back into fact, we sit and watch the Atlas go dark" (296). The Atlas Mountains of Algeria lie in the background, but the deeper meaning depends on recognizing Russell's earlier sentence-in-the-air.

The novel's final sentence demonstrates that Russell has finally found it within himself to take Candace's advice. At a metafictional level, that is, the novel we experience as *Generosity* constitutes "what happens just before" the only sentence that came easily to its narrator. Russell was again hampered by his fear of damaging actual people through the telling, but Candace convinced him to "change it all, slightly, so no one gets hurt" (153). *Generosity* is not just a tale about how an unusually joyful young woman changes the lives of others and then is transformed herself by letting her genome go public. On a more profound level, it is a meditation on the parallel processes of artistic and scientific creation. Thassa thinks of making her videos as "compositing," splicing together various bits of footage to make unpredictable new wholes, and the term applies just as well to the work Russell produces, the book we experience. This is why the text's full title is *Generosity: An Enhancement*. One expects the subtitle to read "A Novel," as it does with *Orfeo*, but Powers is offering an additional hint about his tale. The novel's conclusion reveals that "Miss Generosity" is an enhancement of Thassa's biological inheritance: she might have an unusual genetic predisposition toward joyfulness, but it is not irrepressible. The implication is that there remains a choice, a constant decisiveness about living generously, that it cannot simply be chalked up to how she is "programmed." The subtitle might also point to how the text seeks to enhance the capabilities of the novel as a genre, that it represents the outcome of a process of conscious artificial selection. It is simultaneously a story and a record of that story's composition, an exposition of its raison d'être.

Why does it matter for readers to grasp this additional layer? Because for all *Generosity*'s sophistication, Powers is not trying to write a "perfect" novel, a work complete unto itself. As he told Jian Sun in a 2013 interview about the "stereoscopic" nature of his fiction, "What I want is for the reader to realize that storytelling is not a completable process, is not a finite fixed thing or perfectible thing, that it is an eternal process that has to undo itself in order to keep going forward" (338). Powers has long refused to buy into the binary assumption that he must write either straight mimetic realism or above-it-all postmodern metafiction, insisting that he was formed equally by Hardy and Joyce and that each tradition needs the other.[10] Instead, Toon Staes rightly explains that Powers depends on a habit of "denarration," a term he borrows from Brian Richardson to describe a process of negation by which a narrator undercuts his established plot, not in order to preclude the reader's immersion in the narrative world, but to provide a *simultaneous* focus on its constructedness (50). As he put it to Sun, Powers wants to create "fiction that presents a real story and at the same time temporarily holds that story in abeyance or subverts that story or changes that story to make the reader aware that they are participating in the creation of a representation" (338). In *Generosity*, then, Powers is not using his protagonist's hidden role as narrator merely to wink at the most sophisticated readers but to ask in all earnestness what good fiction can do in a time defined by such enormously pressing "real" problems. He is directly confronting Russell's fear that storytelling might no longer matter when it seems to have been "lapped" by engineering.

Powers is far too subtle an artist to provide a single pithy answer to this concern, but we should especially note a cryptic paragraph that appears very late in the novel, right after Kurton declares that meaning is possible only now that humanity is sequencing genomes. Set apart with glyphs, it stands without further explanation: "Saint Augustine, the old Berber, once wrote, *Factus est Deus homo ut homo fieret Deus*: God became man so that man might become God. He also said, even more popularly, *Dilige et quod vis fac*: Love, and do as you wish. But that was before our abilities so far outstripped our love" (252). In other words, Powers answers his question about story's value with one about incarnation, progress, and love. But how does Augustine's observation apply? What kind of love is sufficient to a world swiftly moving toward the possibilities imagined in *Gattaca*? That film's challenge is set up by the protagonist's voice-over: "They used to say that a child conceived in love has a greater chance of happiness. They don't say that anymore." What about in *Generosity*, wherein one of the most joyful characters ever to grace the pages of American literature slowly loses her soul? Only by understanding Russell's role as a fictional *fiction writer* can we fully grasp the significance of Thassa's inner joy and its gradual eradication, the entire novel's encouragement to "love, and do as you wish."

As with *The Echo Maker*, everything comes down to a suicide attempt. At the end of the novel, Russell has rescued Thassa from the gawkers, the endless stream of people who want something from her they can barely define. The former teacher and pupil drive toward Michigan and then Canada, hoping to get her out of the country, but just past Notre Dame, he has an "epiphany," realizing "why he could never in his life or anytime thereafter write fiction: he's crushed under the unbearable burden of a plot." The problem, he explains to himself, is that "realism—the whole threadbare patch job of consoling conventions—is like one of those painkillers that gets you addicted without helping anything. In reality, a million things happen all at once for no good reason" (273). Realism seems insufficient: purpose, meaning, and truth are fleeting phenomena, mirages serving short-term evolutionary ends. Or so Russell the character convinces himself, especially as he listens to Thassa confess that the "nonfiction" she composed for his class was far more "creative" than he thought, that her moving portraits of everyday Chicagoans were "assembled from some separate parts" (279). Of course, this does not make them lies, no more than Gerald's composite case studies in *The Echo Maker* were lies: they are "performance, in place of the real," "what could have happened," "what might have" (282). Even more troublingly, her work for the creative nonfiction class was "not her only fiction. She has been authoring something else. How high is her real emotional set point, by nature? How happy is she, *really*?" (283). The revelation is not that her joyful demeanor was a facade but that while it overcame the murders of her parents, it was not indestructible. Genetic inheritance matters enormously, but given enough of the wrong environment, even Miss Generosity ends up swallowing a bottle of pills.

What also matters, beyond genetic predisposition and environmental conditions, is agency, the capacity to compose one's own story. That is why when Russell finds Thassa collapsed in the hotel room, what happens next is so significant for understanding genetic metafiction's potential. He realizes that she has been living out Hamlet's biting rejoinder to his mother as if it were the key to life: "*Assume a virtue, if you have it not*. A little creativity with the facts. Lie, if it keeps you alive" (285, emphasis in original). Thassa has been living by myth, and her suicide attempt forces Russell to make a choice: Will he join her? Rushing to her side, Russell "holds his finger underneath her nose" and feels only "the vacuum of deep space" (288). This is critical: she is not breathing. The subsequent paragraphs, then, represent the teacher's decision to learn from his student:

> He scrambles to his feet and heads for the door, the phone, the bathroom faucet, all at once. He hears a voice tell him that he needs to get her to throw up. He can't figure out how he's supposed to do that. He sits down on the floor, shaking, clouded, and adrift. And in that instant of annihilation, art at last overtakes him, and he writes.
>
> He can rescind this. He works his way back to the bed, pauses his hand under her nose again: the slightest, world-battering typhoon. (288–89)

To be clear, Russell's narration does not indicate that there is only a brief pause in Thassa's respiration and then it resumes. The reader finds here only what Russell *chooses* to write, to rescind. Awaiting the paramedics, he "prays to something he doesn't believe in" (289), even "promises God that if she lives, he'll become another person" because "he can do nothing for her but revise" (290). That is, Russell the narrator is explaining how he finds it within himself to undertake fiction, an art of which Russell the character thought himself incapable. In effect, a capacity for story takes genetics beyond the vicissitudes of inheritance and the onslaught of environment to a triangular relationship that includes these factors as well as genuine agency.

Russell's choice is directly contrary to both his former class's antipathy toward revision and Kurton's impatience for fiction. When Russell first begins working with his writing students, he realizes there is "no way they're buying perpetual revision. Half of them don't even believe in the Shift key" (82). Yet "perpetual revision" is Powers's key for not only fiction but biology; it is the only way *anything* grows into itself. Russell's young artists want to rush forward, always expanding but never pruning or deepening, and Kurton has much the same problem, only it is far more ingrained. Note how fully the geneticist's thoughts demonstrate his antipathy toward individual subjectivity: "The whole grandiose idea that life's meaning plays out in individual negotiations makes the scientist wince. Intimate consciousness, domestic tranquility, self-making: Kurton considers them all blatant distractions from the true explosion in human capability. Fiction seems at best willfully naïve. Too many soul-searchers wandering head-down through too many self-created crises, while all about them, the race is changing the universe" (229). Kurton goes on to express his particular annoyance with Albert Camus, whom he takes as illustrating fiction's habit of mistaking correlation for causation (but whom Powers relies on for the novel's epigraph, which translates as "True generosity toward the future means giving our all to the present"). What again grows clear are the consequences of letting methodological naturalism be consumed by the metaphysical variety, and thereby equating fiction with untruth. Kurton regards his own background as "irrelevant," conflating his role as scientific interpreter with maintaining the integrity of laboratory data. His doctrine is that "the double-blind study frees human history from the trap of bias and sets it loose in a place beyond personality," so he dreams of a new kind of collaborative "post-genomic fiction" that "shakes free of the prejudices of any individual maker" (230).[11] I have been arguing that in its attention to interpretation and agency, genetics—at least the personal genomics and garage biohacking that Powers's novels examine—moves far closer to genetic fiction than it might appear. Ultimately, Kurton embodies the opposite ambition, that story might be assimilated by the exactitude of laboratory formulations.[12]

Despite its relative brevity, *Generosity* does not surrender its insights easily, especially because of the subtlety with which its narrator's identity gradually emerges. However, the only alternative is to buy Schaefer's argument that in fact Thassa succeeds in committing suicide, and that it is only the novel's final scene that Russell composes as a last-minute redactor, a man who cannot face what he has enabled and is finally writing fiction only in the worst sense of the term, as a *denial* of reality.[13] This is akin to the mistake of taking the fullness of Kurton's character as evidence that Powers actually shares the character's scientism.[14] Perhaps not coincidentally, such misreadings are more difficult in the case of *Orfeo*, wherein the principal scientist's personality and conviction are more like those of Thassa and Russell. Els has his weaknesses, but he is a Renaissance man who straddles the worlds of music and genetics and retains a consistent suspicion of false binaries. In *Generosity*, the division is over the significance of genomics for the direction of history: "Preservers against revisers, sufficers against maximizers, those who think the book is coming apart versus those who think it's coming together" (239). Then, just as Powers refuses to choose sides in predicting the future of genomics, he stands behind Els's challenges to other binaries in *Orfeo*. As a musician, Els is constantly seeking ways of drawing together high and low culture, the joys of entertainment and of nuance. Looking back at his musical career, "he knew no earthly reason why he should have to choose. Yet he now saw—crazy late—just how the pecking order worked. The high-concept men got all the performances. The twelve-tone formalists got all the cachet. . . . And so the choice came clear to Els: radiant versus rigorous, methodical versus moving" (96). *Orfeo* is dedicated to deconstructing this binary not just in terms of music theory but also as it bears on genetic engineering and fictional form. By the novel's conclusion, Els "comes clean" (346) that he has always valued beauty, but he has spent a great deal of his life "fleeing from narrative," and he only gradually discovers, "to his surprise, that it might not be too late to embrace the kind of storytelling that the world craved" (262). Discomfited to find pub songs creating "more communal pleasure in three minutes than [his] music had created in thirty years," he finds the courage to make art he actually likes, even if it will be hidden from most eyes, deep within the DNA of his bacteria (227).[15]

The central binary that *Orfeo* contravenes is even more understated than radiance versus rigor, beauty versus function, or innovation versus formula. Long after fame is past, Els spends his days as an adjunct at a small college in Pennsylvania. Most of his students want only technical guidance, but Els advises the best of them, "*Do not invent anything; simply discover it*" (309). Only one or two grasp the distinction, but the whole novel asks readers to do so, much as Powers's short story "Genie" suggests that "the world's fullness is not made but found." Powers's meaning is also illuminated by the paradoxes embraced in *The Echo Maker*, the calls to sense logic's dependence on feeling, to save one's soul by losing it. These maxims point to a reality in which truth is less "out there" than within, one that depends on scales falling from one's eyes. This interpretive transformation cannot be forced, only allowed, and its value is what makes Els regret the "omnipotence" allowed by software-enabled composition. While marveling at the near divinity conferred by such tools, he "yearn[s] for the clumsy, freighted flights of earthly instruments." Indeed the verisimilitude of digitized music sends him back to "the exhausted vocabularies of the old masters, looking for lost clues, trying to work out how they'd managed, once, to twist the viscera and swell whatever it was in humans that imagined it was a soul. Some part of him could not help believing that the key to re-enchantment still lay in walking backward into the future" (97). Like Russell riding backward in a Chicago train at the beginning of *Generosity*, this strange orientation recurs with minor variations throughout *Orfeo*. Els hopes "to use regions of cycling pitch groups to create forward motion without resorting to the clichés of standard harmonic expectation, but without falling into serialism's dead formality" (139). In one of the novel's most exquisite passages, he reflects, "There's no fixed tonality, but the sequence still propels the listener's ear through a gauntlet of expectation and surprise. The method feels like a way forward, a middle path between romantic indulgence and sterile algorithms, between the grip of the past and the cult of progress" (187). These are Powers's partial answers about how to make living art in the twenty-first century. Printed and framed as maxims, they might be displayed as appropriately above a lab bench as a piano.

I have not yet pointed out an obvious but very telling element of Powers's style: he loves the present tense. It is a primary tool he uses for keeping his back up against the future. This is fitting, given that his oeuvre regularly calls for a new wonder at the natural and the beautiful even as it admits that these modes can become routes to repression of the past and naïveté about ages to come. Observing a jogger who is tagging MP3 songs for later, seemingly never stopping to hear an entire track, Els asks, "*I hope you have a tag for 'sooner,' too?*" (77). As he laments, "Over almost four decades, people in every North American demographic had lost, on average, somewhere around one-third of their 'sustained focusing interval'" (84). By contrast, *Orfeo* is a slowly unwinding testament to life's brevity: the best music might say, "You're immortal," Els reflects in a tweet, but "immortal means today, maybe tomorrow. A year from now, with crazy luck" (120). Powers repeatedly underlines the transience of life, as when Els advises his students, "Make the music that you need, for need will be over, soon enough. Let your progressions predict time's end and recollect the dead as if they're all still here. Because they are" (322). In guidance recalling Thassa's personal motto—"I try my best to decide no more than God" (*Generosity* 9)—Els remarks, "Listen, and decide nothing. Listen for now, for soon enough there'll be listening no longer" (362). There are many more such reminders sprinkled across *Orfeo*, but they culminate in the novel's fulfillment of its mythic title. Els has made his divine music, and as he expected, his audience has proven unready. He can no longer live in the past or the future, so on the novel's final page, he plunges out the door of his daughter's home, presumably drawing the fire of the special forces surrounding her house, determined to keep any biological weapons he might wield from spreading farther.

This tragic conclusion makes it easy to overlook that it is not just Els, a fictional character like Barbara or Thassa, who is effectively committing suicide but, in a real sense, the *reader*. In *Orfeo*'s last ten pages, the novel shifts from the third person to the second, so that it is not just Els breaking through the years of estrangement from his daughter and then rushing out into the night with vase upheld as if it were a flask full of biological weaponry—it is *you*. The novel's final sentences read, "You'll keep moving, vivace, as far as you can get, your bud vial high, like a conductor readying his baton to cue something luckier than anyone supposes. Downbeat of a little infinity. And at last you will hear how this piece goes" (369). As with *Generosity*, the novel's last words teach readers how to make meaning of everything preceding them. My conviction is that *Orfeo* is whispering once more that finding the soul requires losing it. Long before the final page, Powers's Whitmanesque protagonist tweets a promise to "bequeath myself to the dirt" (311), to end up "under your boot soles" (315), to eventually "filter and fibre your blood" (334). Els is *becoming* his music, the soulful notes inscribed within his bacteria's DNA. Like *Orfeo* itself, Els's tweets will spread unpredictably, shrinking the distance between the microscopic and the human, the human and the telescopic. If they contain any message, it is that everything across this spectrum is knitted together, just as past and future are bound up within the ever-evolving present. The world does not wait for artists and scientists to make something entirely new under the sun but to rediscover the riches already there, to cast them in new forms and for new purposes.[16]

Powers's habit of taking apart a character he has made, not to express cynicism about the possibility of meaning, but to invite others to make their own art, sheds light on Els's composition earlier in the novel. Near its conclusion, "the pianist broke off in the middle of an ostinato, stood up, and left the stage"; eventually, "the horn, too, grew forgetful; he stood and wandered, climbed down the front of the stage and into the audience" (160). Rather than reaching a triumphant crescendo, Powers's musician narrator trails off into darkness, creating space for something new to arise. Stephen J. Burn observes that such movements have long typified Powers's aesthetic: "Having established a haven for the reader through the mimetic foundations of his fiction, Powers ultimately challenges the realist illusion by revealing its constructed nature" (xxxi). This is also what happens at the end of *Generosity*, where Powers slowly removes the scene's components: a book disappears, then a camera, a menu, eventually even a character. All that is left, Russell's narration avers, is Thassa, "my invented friend, just as I conceived her, still uncrushed by the collective need for happier endings. All writing is rewriting." The effect is to once again draw together traditional definitions of literature with genetic writing: all genetic engineering is reengineering. The genome itself is a novel full of details that are set in advance yet still await interpretation, new "readings" every time through. Russell was afraid of composing fiction precisely because he could not control those meanings. The same potential for both good and ill characterizes personal genome testing, gene editing, and other forms of synthetic biology. Yet if at the end of Russell's story, Thassa can smile "as if to claim once more that fate has no power over anything crucial," so too might readers hope for a world in which genetic predisposition is real but far from all-determining or immobilizing (295).

Back in *Generosity*'s classroom, long before the novel dissolves into its constitutive elements, Powers uses Russell's creative writing textbook to remind readers that "denouement *doesn't mean tying up all your loose ends. Quite literally, it's French for untying*" (112, emphasis in original). This assertion is fascinating for two reasons. First, the word *religion* is derived not just from the Latin *religio*, which indicates reverence for the divine or sacred, but also from *religare*, which denotes the act of binding or holding together. Juxtaposing the second definition against Powers's elucidation of *denouement* suggests how he strives for a fiction that operates oppositely to dogma, a hybridized creative nonfiction in which metanarratival layers are not just aesthetic glosses but critical elements for making and then unmaking mythology. Second, in his 2006 essay on genetic testing, just after Powers hinges his response on whether a given examination "increase[s] a patient's ability to write her own life," he adds, "It pays to keep in mind that the *denouement* is not the tying up: the word means, quite literally, 'untying'" (46). Thus Russell's textbook gives the same advice about writing and interpreting novels that Powers himself gave about interpreting a genome. The upshot is that whether examined genetically or fictionally, human beings are neither meaningless products of fate nor all-transcending entities who can become anything we might imagine. We are possessed, for a time, of a hybrid status, a predisposed agency that depends on innumerable environmental influences. In *Generosity* and *Orfeo*, as with the best tales of genetic metafiction—or "creative nonfiction"—genes are not predictors of an absolute destiny but the prerequisite instruments of multiple future compositions, each of which plays out in its own unforeseen manner.

Coda
Arrival

As *Generosity* begins, on a train heading to his first "creative nonfiction" class, Russell Stone sees a dark-haired boy "with a secret quickening in his hands. Something yellow floats on the back of the boy's curled fist. His two knuckles pin a goldfinch by the ankles. The boy quiets the bird, caressing in a foreign tongue" (5). Over the next three hundred pages, the moment is easily forgotten, but the image recurs near the novel's final page, this time in what seems a more natural setting. Just before meeting Thassa in Tunisia, Tonia Schiff glances across a dusty street and sees "a shirtless boy sitting on a three-legged stool talking to a yellow bird he pins gently between two fingers" (291–92). The match cut quietly encapsulates Powers's evolutionary aesthetic: there is both repetition and difference in the scene's transposition from inner-city Chicago to a remote African town. On one hand, the novel narrates real linear movement, from Thassa's idealism to her disillusionment, and from Russell's timidity to his newfound courage; on the other hand, it acknowledges a sense of endless recurrence, from motifs like the boy with his finch to events like the birth of a happy, brown-skinned girl created via Thassa's harvested eggs. *Generosity*'s driving philosophy represents the best of genetic metafiction, complete with all its "creative nonfictionality": "Say it has happened already, just the way it will" (80). As Jay Clayton saw at the beginning of the century, genome time untethers easy assumptions about linear causality and straightforward determinism, pulling material existence into a simultaneous awareness of circularity and simultaneity, the interdependence of all things.

In this book's later chapters, we saw how individual and communal memories can be simultaneously fictional and truthful, full of both improbable repetitions and subtle variations. Acknowledging the dually inherited and constructed nature of one's reality does not devalue it; seeing the seams in our stories and the DNA shaping our predispositions should make life more precious, not less. For this book, that has meant studying stories about genetics that rely on varying combinations of fantasy, realism, and metafiction. My proposal has been that when understood according to their unique modes, these science fictions serve as more insightful and modest witnesses of our biotechnological present than much popular rhetoric about genetics, including that employed by some scientists. On its surface, *Editing the Soul* is a map by which to understand the ongoing mutations of short stories, novels, graphic narratives, television shows, and films about new possibilities for genetic knowledge and intervention, grasping their distinct deployments of analogy, metaphor, and real-world biological references. Diving deeper, the book invests especially in these narratives' growing readiness to break the fourth wall and reflect on their own capacities. Genetic metafiction invites readers and viewers into greater consciousness of human transience, exposing the parallel vocabularies of denial and inevitability by which twenty-first-century lives too easily excuse conformity and inaction.

This book is fueled by a postsecular resistance to the ideology of technotranscendence, whether it might entice us toward genetic determinism or repulse us into genetic dismissivism. Avoiding both sides of the precipice, we must take seriously humanity's persistent hunger for personal and communal meaning making, one sometimes satisfied by sources deemed "religious," other times by art, entertainment, sports, politics, and other cultural activity labeled "secular." Not every reader or viewer needs to find biblical references around every corner, no more so than we should ignore the insights of those who do. Instead, we must attend to the processes by which humanity's most promising and threatening discoveries and inventions can be obscured by absolutism, whether of the fundamentalist or scientistic variety. From this perspective, the possibilities of personal genome testing, gene editing, and various forms of synthetic biology are like other new powers that have entered the modern world. They require honest and nuanced representation in advance, thoughtful and transparent reports as they become familiar, and sober and public assessment as they evolve into new forms. Of course, such ideals are not easily achieved when there are billions of dollars at stake in biomedicine, when treatment recipients and biomedical corporations evaluate trials according to different criteria, and when political discourse devolves to the point that the most basic shared facts about science and history cannot be assumed. But that is why we need a wide variety of richly particular narratives, so that we may look and feel beyond the lure of easy generalizations and abstractions to the very personal stakes of new knowledge and therapies.

I close this study with one last film, *Arrival* (2016), which exemplifies the power of story to reflect on the processes by which we make meaning from data. My choice of this concluding text might seem curious in that Denis Villeneuve's film does not feature direct representations of genetics. However, it is deeply concerned with epistemology and with the relationship between determinism and agency, and it offers a rich opportunity to show how this book's methodology and findings reach toward even broader questions about reproduction, ecology, and humanity's present and future. Like the Ted Chiang short story that inspired it, this alien-contact story asks how, when, and with whom we make babies, decisions undergirding many bioethical debates. In the process, *Arrival* cultivates a postsecular challenge to absolutism closely akin to that found in the preceding chapters. I have argued throughout this volume that no matter how much genetic information we have at our disposal, such data still points toward uncertain timeframes and sometimes difficult interpretive choices. *Arrival* is an intricate thought experiment that presses this situation to its limits, asking how to respond to even relatively certain knowledge of future events, and in that sense, it deploys questions that are inextricable from genetic testing and editing. Indeed, the capacity to see around the edges of time—acquired by the film's main character—is an enlargement of the knowledge many seek in spitting into a tube and then mailing it off to be translated into nucleotides and then statistical likelihoods of developing various conditions. In seeking to interpret the heptapods' unconventional screen language, Louise Banks can be understood to stand not only for cinema audiences generally but especially for those who peer into the futures that are partially revealed by their genomes.

Arrival's premise is simple: twelve alien ships suddenly park themselves above various locations distributed around the globe, including a nondescript field in Montana, and someone has to figure out how to communicate with the wonderfully nonanthropomorphic occupants. In the United States, the military takes charge, ascending into the cavernous ship when a cuboid tunnel opens on its underside once every eighteen hours. After the failure of previous interlocutors to handle the spatial and cultural reorientations involved, the army recruits two civilians to lead the encounters in the viewing chamber at the shaft's apex, a space that is very much a metacinematic referent to the theater in which the film is projected. There is some mild disciplinary squabbling between this linguist, Louise, and her physicist colleague, Ian Donnelly, but his scientific hubris soon gives way to respect for her expertise in a field he barely knew existed. Eventually, their shared task of interpreting the heptapods and their logograms births a romance. At some point after the aliens depart, Louise and Ian have a daughter, and by referring to encounters with her as a child and a young teenager, Louise is able to make the film's most critical intuitive leaps.

Notwithstanding the appearance of spaceships from distant galaxies, this film's real novum is its representation of time. Audiences initially assume that Louise's conversations with her daughter, spliced with increased frequency into scenes following the shipboard sessions, are memories. Instead, these moments lie well into Louise's future; they are visions without which the child they feature could not have been born. The film's big reveal is that the heptapods' "weapon" is not the destructive tool anticipated by national governments but a universal language that remaps the brains of its users so that they perceive time not just sequentially but also simultaneously.[1] This capacity is something like that of the physicist to analyze light as both particle and wave, or as Chiang's story notes, that of viewers of a classic optical illusion to shift between seeing the face of a young woman and an older one.[2] The two ways of seeing are compatible but distinct; they reveal congruent information about the same object via different subjective angles and imaginative expectations. Similarly, by learning the aliens' written language, Louise develops the ability to experience time not only in terms of causation but also with a more holistic eye to purpose. She cannot experience her daughter as future possibility and present reality in the same instant, but she shifts between these points of view at increasingly frequent intervals, just like the steadily progressing student of a new language who eventually starts *thinking* in that new language, not just translating thoughts in their first language to the new one. Louise's skill is neither a one-way conversion from one way of seeing to another nor a fusion of opposite ways of being but the *expansion* of her normal way of experiencing time via another.

This ability to alternate between sequential and simultaneous approaches is closely akin to the perspectival agility we need in our postgenomic age. Louise learns to embrace a far more specific knowledge of future circumstances than anything genetic data can provide—certainly a far more detailed prophecy than the probabilistic litany of *Gattaca*'s birth scene—yet she does not withdraw into fatalistic apathy. Like the heptapods, she continues acting in her immediate present, but it is backlit by a more distant awareness. She remains an enstoried being living with the limited knowledge of a causal universe, yet she also can glimpse the immeasurability, the ultimate fictionality, of seemingly commonsensical divisions among past, present, and future. (Ponder this: When exactly is the present? *Right now, obvio*—but wait, how long was that "now"? Or consider Zeno's paradox: At any point in a thrown ball's flight toward a wall, it can always cross half the remaining distance. By implication, it should never actually arrive.) In short, at some point the language we use to describe time reveals itself as more pragmatic than exact. This is what Louise's ability demonstrates in larger, more literal terms than nonphysicists generally use. We all have some ability to conceive of our lives as a mere instant within an inconceivably larger narrative, but Louise's constructions of future events are far more specific than the rest of us achieve. Here the story partakes of fantasy, of course, but it does so quite consciously and purposefully.

Because of her daughter's future persistence in asking about "that technical term" describing two parties making a deal and both benefiting, Louise is able, back in the moment of crisis, to envision the possibility of a "win-win," a "non–zero sum game" for international relations. It is this temporal suppleness that allows her to grasp the heptapods' nonlinear logic, which includes their indifference about the question of which occurs first, their act to save humanity "now" or humanity's to save them "three thousand years from now." Reaching beyond the nationalism and militarism surrounding her, Louise makes a phone call that convinces a Chinese leader to stand down rather than attacking. She has General Shang's cell number only because he *will* show it to her in the future, intuiting but not entirely comprehending the importance that he do so. She is able to persuade him to trust a complete stranger because she *will* know what only he could have shared, his wife's dying words. Through this time-bending feat, humanity rescinds its aggression against the heptapods, averting a mistake of literally cosmic proportions. Implicitly, *Arrival* counsels that humanity's future depends not on technological advances or bureaucratic negotiation but on the capacities to imagine the interdependence of nations and species and to leaps of faith rather than defaulting to a fearful protectionism.

We should directly acknowledge that the film's diegetic chronology is impossibly circular, at least according to the logic of classical physics. As James Gleick's review astutely observes, "If Louise prevents the war and saves the world by phoning Shang, surely she will remember that at the celebratory party." However, the sense of contradiction here depends on a traditional notion of causality. Human beings normally imagine the future as inaccessible, but quantum physics suggests otherwise, revealing our separation of *this moment* from previous and subsequent ones as a necessary myth. As this book has often observed, we rely on such fictions to make meaning of our lives, and whether they are immaterial or unprovable does not render them valueless or untrue. On the contrary, we cannot do without them; they are not science, but they make science matter. This is why it is so significant that the film represents time not only as circular but as circular *in addition to* linear. For all the linguistic rings that appear in its mise-en-scène, the ultimate shape of *Arrival*'s narrative—and those who have read this far will not be surprised—is spiral. As with the golden ratio of Cosima's tattoo in *Orphan Black* and the spiraling narrative shape of *Generosity*, the film embodies Peter Brooks's repetition with difference. The logograms of *Arrival* inscribe their ends in their beginnings, but they reach beyond mere duplication.

Louise says in the introductory narration that she is "not sure [she] believe[s] in beginnings and endings," but she does not dismiss the notion of progression, nor does the film's structure. She addresses these words, like those of the film's conclusion, *to her daughter*. The overdubbed narration positions the film as a speech act with eventual consequences for both its diegetic audience (her daughter) and its extradiegetic one (us). From a sufficiently abstract perspective, "later" might be a relativistic concept, but it remains necessary. In another synthesis of forward and backward movement, the protagonist names her daughter with a palindrome, Hannah (much like Atwood's choice of the title *MaddAddam*). *Arrival*'s commitment to oscillation, to illuminating the significance of both repetition and forward momentum, extends even to the film's sound design and costuming. The film uses Max Richter's achingly beautiful melody "On the Nature of Daylight" in both its rhyming opening and closing scenes, each of which begin with a slow tilt of the camera down the darkened ceiling of Louise's living room to silhouette her against another figural screen, her home's windows onto an unidentified body of water. However, these sequences diverge with the appearance of a new character in the conclusion, Ian, and his invitation to "make a baby," yielding both repetition and difference. Likewise, during Louise's climactic conversation with General Shang, she wears distinctly spiral earrings, the film's most overt acknowledgment of its marriage between sequence and simultaneity, linearity and circularity. Thematically and formally, *Arrival* echoes the narratival self-reflexivity that builds across this volume, a metafictional awareness that humanity's story involves both forward velocity and curvature, both progression and return, much like the curling hospital hallway Louise navigates upon her daughter's death.

To flesh out this paradox only a touch further, consider two essential shapes that emerge from both this film and *Editing the Soul*. These are the forms in which the alien spaceships appear onscreen: initially from the front, as an enormous egg, and later from the side, as a giant lens (figs. 18 and 19). The ships' gray ovoid design is a telling revision of the pitch-black, rectangular monoliths in *2001: A Space Odyssey* (1968), one of several major first-contact films that *Arrival* references. Whereas that classic culminates with Dave Bowman's rebirth in the form of a "star child" gazing down on a similarly sized Earth, *Arrival* offers a more grounded, nuanced vision of reproduction. This is signaled visually in the highly regimented, patiently edited sequences in which we watch the humans ensconce themselves within plastic hazmat suits, looking for all the world like giant condoms, and then penetrate the alien spaceship's enormous egg, all according to a fixed cycle. I will refrain from any extended psychoanalytic interpretation—there is slippage here between egg and womb

Figs. 18–19 The alien ship viewed as an egg (fig. 18) and the alien ship viewed as a lens (fig. 19) in *Arrival* (2016).

Coda

and between sperm and phallus—but these images are an evocative context in which to understand Louise's choice to remove her suit during her third session with the heptapods. Claiming that "they need to see me," she discerns that personal vulnerability is essential to any true form of community, that the prophylactic measures on which her species has depended thus far have created too great a barrier against trust. She has to bridge the gap between the brilliant white of the screen and the shadows in which the other characters remain; she has to enter the gray area, the space of uncertainty, of greatest risk.

Crucially, that choice threatens not only the military's hierarchical, sequential decision making but also various characters' performances of masculinity. Louise decides to rely on the good faith of the beings beyond the antechamber screen as well as the fact that a small bird carried into the ship has shown no ill effects from its air. Like the finch Powers uses to accentuate the spiraling evolution of *Generosity*'s structure, the canary in this inverted coal mine inspires Louise to set aside her bodily shield. Seeing no harmful consequences, Ian soon follows suit. Despite the barely contained boyish glee and clumsiness with which he boarded the alien vessel, he stands at the less extreme end on the film's spectrum of hypermasculine defensiveness. Colonel Weber also displays a self-restraint that goes against type, risking his reputation on Louise's slow work toward a shared vocabulary. But this is difficult; as he puts it, "Everything you do in there, I have to explain to a room full of *men* whose first and last question is, 'How can this be used against us?'" The film reverses a typically female-gendered hysteria, expressing it even further through Captain Marks, who, along with several other male soldiers, fears his superiors are acting too passively. One of the first questions Louise must answer upon reaching the hastily built army base near the ship is whether she is *pregnant*; not coincidentally, then, when Louise and then Ian suddenly remove their suits, Marks's repeated, panicked question to his superiors is "Should we *abort*?" Such ejaculations align the military's defensiveness toward the alien ship with deeper anxieties about the female body, which are then stoked (appropriately) by Marks's exposure to right-wing fearmongering via online media. The result is an attempt to destroy the ship by smuggling aboard explosives, a simultaneous rape-homicide that partly duplicates the scene in *Contact* (1997) in which a fundamentalist preacher blows up a machine also designed to transport human beings into alien spaces. Like that film and the Carl Sagan novel it adapted, *Arrival* quietly exposes a hypermasculine inability to undertake relationships with strangers in terms other than sexual, biopolitical, or religious domination.[3]

Editing the Soul has shown many times how these lexicons continue to galvanize one another. Thomas Kurton talks to the camera about the possibilities of bioengineered life extension, but he is already appraising Tonia Schiff's hair, preparing to seduce her. The Rev sells transcendence to his burgeoning congregation, but behind the scenes, he plays video games in which he rapes and executes women and then ends up dead in a "gentlemen's club" after trying to have his son Zeb murdered. Allison Mann's father dreams of remaking the world in his own image by endlessly cloning his daughter. Marcus Chalfen only modifies oncomice genes, but his cheerfully subservient wife grasps that this is just one more step in his gradual self-divinization, a eugenic trajectory already established by his mentor, Dr. Sick. Henrik Johanssen imagines that a "war for creation" justifies treating women as brood mares, including his own daughter, while Aldous Leekie's treatment of women as property stems from the shinier cult of Neolution and its corporate ideology. We could continue working through this book's primary texts, but it should be overwhelmingly evident that genetic fiction regularly exposes the congruity between the hierarchies enabling sexism and militarism and those promoting scientism and fundamentalism. Similarly, *Arrival* demonstrates how an obsession with power yields suspicion and self-destruction but counters it with the way of the open hand embodied in Louise's acts of self-exposure. Ultimately these lead to her invitation beyond the screen's mediation: stepping into a pod that detaches from and then returns to the ship's side, as if offering an alternative, asexual mode of engagement, she finds herself surrounded by the aliens' glowing atmosphere and discovers how little of them she could see previously. It is in this final session that she fully understands the mixed blessing she has been given, one very much like that of many new genetic technologies. Rather than abandoning any concept of the present, though, she can now reshape it via greater awareness of its wholeness with past and future.

If the alien ship's egg shape emphasizes *Arrival*'s concern with reproduction, its lens shape (clear only in the film's second hour) brings us back to epistemology, specifically its advocacy of subjectivity. For Villeneuve's film, we see things not as they absolutely are but in terms of other things, and *this can be a good thing*. The lens metaphor is signaled as early as the moment when Colonel Weber first appears in Louise's office and we see her computer monitor's tall, thin, angular shape profiled in the foreground. We cannot avoid the interpretive uncertainty involved in storytelling, *Arrival* suggests, nor should we. In chapter 1, I noted how Michael Bérubé's 1995 memoir reflected on his young family's experiences with Down syndrome and the need to measure value in terms beyond financial cost or usefulness. In 2016, Bérubé published a companion volume that contemplates his son Jamie's experiences now that he is a twentysomething, a young man whose condition is simply part of his unique personhood. Lamenting our culture's proclivity to turn impairments into disabilities, especially through overuse of disease rhetoric, Bérubé rejects the assumption that one's background of experience is only a negative limitation. Rather than obsessing over the capacity to assess reality fairly, as if completely unbiased neutrality must exist and all questions should yield to laboratory tests, Bérubé turns to the hermeneutic of Hans-Georg Gadamer. As Bérubé explains, Gadamer contends that "there are forms of 'pre-judgment' that do not necessarily lock one into a narrow dogmatism about that subject"; in fact, "there are plenty of people whose ability to argue a point is *enhanced* by their intimate personal knowledge of what they are talking about" (85, emphasis in original). This awareness is invaluable for grasping how *Arrival*'s lens imagery invites postsecular thinking.

Superficially, Villeneuve's film features no more obvious gestures toward religion than it does genetics, but just as it is deeply concerned with reproduction and gender, it also scrutinizes the relationship between knowledge and transcendence. There are no verbal allusions to Jesus Christ as in *Y: The Last Man*, nor is there Hebrew vocabulary or overt messiah figures as in Le Guin's "Nine Lives" or Butler's *Xenogenesis*. However, when one ponders the choices faced by Louise, they are indisputably Marian; indeed, the position in which she finds herself is remarkably like that of Butler's Lilith. Especially through voice-overs in the opening and closing scenes, the film functions as a modern-day, alien-mediated annunciation. Following but also adjusting the pattern Rebekah Sheldon brilliantly traces in *The Child to Come*, *Arrival* places humanity's future not in the hands of a boy born in a manger, not even in those of a teenaged girl who receives no choice about her role, but in those of the girl's *mother*. Drawing together the seemingly eternal present of genome time with theology's vision of *kairos* time, the film aligns egg and lens, reproduction and interpretation. All is contingent on Louise's finding the symmetry between her present situation and her daughter's future inquiries about a non–zero sum game. Ian remarks at one point, "I feel like everything that happens in there comes down to the two of us," and indeed the film turns on Louise's decision to accept the trajectory of their relationship. Physicist though Ian may be, this figurative Joseph can only imagine the significance of the phone call he helps her make. In fact, his most crucial decision is a gut-level response to a simple question: *Do you trust me?* Louise's introductory narration reflects, "We are so bound by time, by its order," but the film simultaneously recognizes the *lack* of knowledge or control involved in its most critical decisions. Unlike the Pentecostal cult that uses the ships' appearance to justify apocalyptic violence, then, it is crucial that Louise embraces uncertainty. As she tells Ian about one of her visions, "I'm not sure it's something I can explain."

Should she try, anyway? For a time, Louise is unconvinced she should share her sense of future events with her partner. According to one of her visions, though, that changes during their daughter's childhood. Hannah tells her mother, "[Daddy] doesn't look at me the same way anymore," and Louise reveals, "It's my fault. I told him something that he wasn't ready to hear. What? Believe it or not, I know something that's going to happen. I can't explain how I know. I just do. And when I told your daddy, he got really mad. And he said I made the wrong choice. It has to do with a really rare disease and it's unstoppable, kinda like you are." How are we to interpret this soliloquy? Is the film proposing that the future already exists in an inviolable form, so that human freedom is merely illusory? I don't think so. The film is not rejecting a linear, causal perspective on time but *supplementing* it with a circular, teleological one. The mistake is to assume that future knowledge enabled by an awareness of simultaneity justifies passivity within one's sequential experiences. This mistake is why Ian struggles to understand how Louise could accept the future revealed by the heptapods' language. What he fails to consider is that even the sharpest vision of what is happening directly in front of one's eyes involves interpretive choices. We see through a glass darkly, whether we look backward *or* forward. We do not have access to cameras and microphones trained on every square foot of the world across all time, and even if we did, we would have to rely on partial knowledge, constructed knowledge. The map cannot be the thing itself: none of us could hold it all in mind simultaneously, so we would have to choose to look harder at some places and less intently at others. *Arrival*, though, suggests that this can be *a good thing*. How we respond to what we think we know matters as much as the information itself.

This is very much our situation with multiple realms of biological knowledge in the early twenty-first century. We are likely to continue uncovering and generating far more information than we are able to translate into meaning. Using this knowledge will remain an inherently subjective process. This is why it is so appropriate that *Arrival* brings us to the edge of a partially known future and then invites us to take over for Louise. We too are blessed and cursed with greater capacity to anticipate future events than earlier generations were. We know, for instance, that our species' demands on others are unsustainable, that our reflexes toward suspicion, self-defense, and even domination are rapidly destroying a long-standing global ecological balance. Yet we dither on, to use one of Kim Stanley Robinson's best words, in large part because so many of us accept the logic of either biological determinism or dismissivism. It matters little whether we decide that climate change is a hoax or that its worst consequences are inevitable; these opposite assumptions cultivate the same unaffordable inaction. Like the best stories about biotechnology, though, *Arrival* points in a more challenging direction. The film is an invitation to confront partially known futures with a mind-set that accepts death's imbrication with life and thus maximizes quality rather than quantity of life. Louise might know more specifics about her daughter's future than the average mother, but how exceptional is she really? All of us who take up the stunning level of responsibility for another life involved in parenthood have every reason to expect such "failure." We will not be able to protect our children forever, no more so than we can our parents or ourselves. The wonder is that we get to play the tickle games and make art together at all.

It is in this spirit that we should hear the most critical sentence in the film, narration that arrives just before the credits. "Despite knowing the journey and where it leads," Louise tells her daughter, "I embrace it, and I welcome every moment of it." This could seem impossibly stoic or naïve. If what lies ahead is death and destruction, some would say that Louise must be fooling herself to think she can "embrace" or "welcome" it. However, the film's narrator reaches this point not in a state of blithe denial but while counting the costs of existence against the glories of a baby's cheek, the joys of playing pretend. She will live with a seemingly divine awareness of temporal simultaneity but within the poignantly human limitations of a sequential narrative. For a time, at least, Louise's choice is infectious, with Ian reversing Vincent-Jerome's decision in *Gattaca* to leave behind his relationship with Irene in favor of fulfilling his dreams as an astronaut. Ian confesses, "I've had my head tilted up to the stars for as long as I can remember. You know what surprised me most? It wasn't meeting them. It was meeting you." That might seem a throwaway line inserted to create some semblance of a happy ending, a distraction from the tragedy and pain. However, it becomes enormously powerful if, unlike Ian, we let it encompass the grief in the story as well as the joy. In the most profound sense, *Arrival* indicates that welcoming genuine love is impossible without also accepting eventual loss. As neurosurgeon Paul Kalanithi repeatedly illustrates in his extraordinary memoir *When Breath Becomes Air*, life and death are bound up with one another.

For some posthumanists, this is a defeatist attitude. For them, we should all be like the vengeful Roy Batty early in *Blade Runner*, a bioengineered being who curses any creator who would deny him "more life." However, this perspective forgets the glory of that character's final scene, in which after reaching down to save Rick Decker with a nail-pierced hand, and after recalling moments "lost in time, like tears in rain," he accepts his ending. *Arrival* also takes advantage of our culture's biblical inheritance, reframing not only the story of Jesus's mother but also an earlier one. Without reducing the film to any one-to-one expression of Jewish or Christian typology, a postsecular lens reveals its allusion to 1 Samuel. Here we learn of Hannah, one among many barren women in the Bible who pleads with God for a child and experiences a miraculous pregnancy. When her hopes are fulfilled, she keeps the promise made in her prayers, and once the baby is weaned, she gives him to the priests at the temple. Samuel grows up to be the last of the judges and first of the prophets, and appropriately, the only ghost to appear in the Hebrew Bible.[4] *Arrival* offers a fascinating twist on this story, with the little girl named Hannah in the film taking on the role of the biblical Samuel. She too becomes not only a ghost but an apparition of the future, one who preexists even her own conception, like Cal/lie of *Middlesex*. She asks Louise to make the same sort of choice faced by the biblical Hannah: Will she embrace Ian and have a baby, knowing that she will not be able to protect her child forever? Louise sees what lies along the path ahead—a break with Ian, Hannah's cancer—and so does the audience. The natural reaction is revulsion, and according to Louise's visions, this is Ian's reaction too. He will accuse her, to be precise, of making the "wrong choice."

However, this is the result of thinking only in sequential, causal terms. In Ian's mind, we are to understand, *Louise accepted the aliens' bargain, and she didn't tell me*. If he takes his marriage and his daughter's death as the price of humanity's survival, though, he confuses the issue at hand. Even from a linear point of view, it is not Hannah's death that is necessary but her *life*, especially her question about non–zero sum games. Assuming that the future Louise sees is actually inescapable in every detail—an assumption the film never questions but also does not explicitly require—the greatest price is the emotional cost of knowing what lies ahead, or merely of believing that one knows. The question Louise really has to answer is whether she can accept the awareness of transience that a holistic approach to time involves, the reality that through a sufficiently wide-angle lens, the choice to make *any* baby is a decision to accept death. As with the biblical Hannah, Louise retains the freedom to break her promise, to interrupt the chronology of events portrayed through her visions. There is nothing that says, for instance, that she could not use her vision of Ian's anger in response to her "wrong choice" to justify hiding her future knowledge more permanently. But we need not grasp after such speculative possibilities. *Arrival*'s core insight is that there can be multiple noncontradictory ways of representing the same reality, that language and subjectivity really are crucial, and that two events that might look inseparable from one angle can still be experienced independently. With this in mind, it is especially pertinent that the film modifies only two main characters' names in adapting Chiang's story. In the film, Gary Donnelly's first name is changed to Ian, which means "gift of God," while the unnamed child in the short story becomes Hannah, which means "grace." These names do not reduce the film to a simplistic allegory, but like the addition of a holistic, circular perspective on time to a linear, sequential one, they tighten the film's embrace of subjectivity, vulnerability, and interdependence.

In sum, *Editing the Soul* suggests that for every binary, there is a third way that is not just a watered-down compromise but a recognition of the dynamic interdependence of the opposing terms. Rather than the parallel paths of militant atheism or religious fundamentalism, which are ultimately reducible to the same road of performed certainty, the postsecular approach demonstrated in this book implies many routes to integrating the known and the unknown. In *Arrival*'s storyworld, I don't know whether all of Louise's visions must play out exactly as she anticipates, but I am moved by her willingness to accept the mourning with the dancing. In our world, I don't know how much genetic biology will be able to reveal about or reshape my future or that of my children, but my plan is to welcome any trustworthy knowledge, neither fleeing nor overestimating it, at least insofar as personal information can be protected from prying eyes. Knowledge becomes meaningful only as it is narrated, and for living, breathing human beings, that means endlessly unpredictable vectors spinning outward from varying sets of initial conditions. It is one thing to be told that the double-blind studies completed to date suggest a statistical likelihood of a particular disease outcome within a given time period. It is another thing to live out those days, to become a character and narrate them, to shape the story's meaning.

This is why we continue to need insightful fictional articulations of biotechnological possibilities, to help us become characters whose stories we might never fully control but that we actively influence. We need the bold analogies of genetic fantasy, the sober cautions of genetic realism, and the self-aware relativity of genetic metafiction. The narrative art of this century is already enormously diverse, but in imagining the meaning of genes and genomes, it consistently invites readers and viewers into nuanced, constructed understandings of personhood and community. In terms of race, sex, gender, nationality, religion, and many other categories of identity, our souls are always more hybrid, dynamic amalgamations than the official forms suggest. Fully embracing our species' embeddedness within the natural world requires accepting our own and others' very real biological predispositions. It means acknowledging that human beings have been editing each other's souls for centuries, and that as the tools grow more awe-inspiring, so do the risks and the opportunities. It is time to rediscover genes not as codes by which to skip to a story's conclusion but as clues for deciphering the many bonds across beginning, middle, and end.

Notes

Introduction

1. CRISPR stands for "clustered regularly interspaced short palindromic repeats."
2. For a powerful half-century-old discussion of this mistake in science fiction more generally, see Le Guin's introduction to *The Left Hand of Darkness*.
3. The term "outsider science fiction" was identified by Andy Sawyer.
4. For an illuminating treatment of this term, see Csicsery-Ronay Jr.
5. The origins of genetic fiction depend on how far back one understands genetics itself to extend. One could argue that it goes back to Darwin, who relied on the notion of "gemmules," or even to the Preformationists, notwithstanding the inaccuracy of their visions of miniature but fully formed organisms.
6. See also Trushell's article, which notes the parallels between the Professor X-Magneto relationship and that between Martin Luther King Jr. and Malcolm X, as well as Ramzi Fawaz's more recent analysis of *X-Men* cultural politics.
7. For historical explanations, one could point to surges in media attention around the 1997 birth announcement of Dolly the sheep or the 2000 press conference about the Human Genome Project's near completion. However, if I had to name a first novel of genetic realism, it would be Powers's *The Gold Bug Variations*.
8. This storyline is dispersed between season one, episodes six through seven, "The Trials" and "Faint Hope."
9. This story arc shares some elements with the shocking 1999 death of Jesse Gelsinger, a teenager who voluntarily underwent a gene therapy trial targeting the metabolic disorder known as ornithine transcarbamylase (OTC) deficiency.
10. For another example of genetic fantasy yielding moments of genetic metafiction, see Cherryh's epic *Regenesis* (2009), a sequel to two Hugo Award–winning tomes, *Cyteen* (1988), and, more distantly, *Downbelow Station* (1981). This *Regenesis* traces the shifts among "Ari 1," "Ari 2," and "Ari 3," Cherryh's nomenclature for three successive clone characters that the novel treats as "edition[s]" of a single human being (498).

Chapter 1

1. *Gattaca* has been featured extensively in work by bioethicists (see Green). For analysis by film scholars, see Clayton's *Charles Dickens in Cyberspace*; Kirby and Gaither; Stacey; and Hughes.
2. To illustrate how rapidly this market is evolving, consider that in summer 2016, Church founded a new company named Genos as a spin-off from Complete Genomics. It offered whole exome sequencing with 75× scanning (roughly fifty to one hundred times the data targeted by 23andMe) for $399–499. But only months later, in April 2017, Genos itself was acquired by NantOmics (for which Church became an advisory board member) and temporarily halted acceptance of new orders. Church has also founded at least ten other biotech companies.
3. Mukherjee provides another helpful analogy for understanding the nature of a gene.

Retelling geneticist Antoine Danchin's parable of the Delphic boat, Mukherjee asks, "How can the boat be the same boat—the riddle runs—if every physical element of the original has been replaced? The answer is that the 'boat' is not made of planks but of the *relationship* between planks." Similarly, he explains, "Individual genes specify individual functions, but the relationship among genes allows physiology" (196).

4. For additional treatments of contemporary genetics accessible to nonspecialists, see Blackford; Moore; Parrington; and Nowotny and Testa.

5. Dickenson provides a wide-ranging and insightful explication of controversies around BiDil (the marketing of which used spurious research to target African American populations), cord blood, vaccinations, and other biomedical innovations. The sometimes counterintuitive nature of her analyses is particularly impressive, as when she explains that the retailing of personal genome tests "is predicated not on genetic determinism but rather on its opposite: an underlying assumption that we are in control of our behavior so that we can alter unhealthy eating or exercise patterns, for example, to counter a genetic predisposition to heart disease." Ironically, "when combined with the 'ideology of wellness,' geneticization means that those who are well can then take credit not just for their superior genes but also for their initiative in counteracting any 'inferior' ones" (15).

6. This awareness suffuses an excellent survey of genetic history, Comfort's *The Science of Human Perfection*. In describing the period between 1958 and 1963 (the "Golden Age of molecular biology"), for instance, Comfort writes, "DNA was alluring, its curvy lines replacing Eden's apple as the iconic depiction of self-knowledge" (206).

7. This term has a complicated history spanning several centuries. For Venter, though, vitalism can be understood as the rough equivalent of "supernaturalism" or merely "religion": it is the enemy of the purely digital, biochemical description of reality that he seeks.

8. I use the term "relative objectivity" as a bridge between the needed caution that science is a human activity, and therefore its discoveries are always approximations open to further revision, and the equally crucial affirmation that its insights grow increasingly reliable as they accrue, and as such become eminently worthy of our trust.

9. Those with interests in genetics and literature should also consult books by Bruce Clarke; Holloway; Lake; Nelkin and Lindee; Roof; and Shea.

10. For a recent, wonderfully concise, but more extensive overview of the emergence and evolution of postsecular theory, see Branch. For an alternative approach to the still unfolding significance of postsecular theory, see the December 2014 special issue of *American Literature*, especially the brief introductory effort by editors Coviello and Hickman to parse the history of postsecular thinking into three distinct phases. For additional major landmarks within this theoretical movement, see especially work by Fessenden and Hungerford.

Chapter 2

1. I am consciously using *self* and *soul* almost interchangeably in this argument, as my postsecular approach makes me interested in blurring the lines between the secular self (the preferred psychological nomenclature) and the religious soul (which more often suggests the domain of theology). While resisting the essentialist assumption that this entity is a static phenomenon, I treat both discourses as efforts to describe the core of personhood.

2. As Mukherjee details, this was actually the second Asilomar conference, but its legacy was so much greater that it is often remembered as the only event.

3. For an encyclopedic survey of Le Guin's fictional dances with Christianity, monotheism, and other religious forms, see Erlich's argument that "the one change Le Guin will not allow is what the nature of Dao excludes: the change to changelessness" (370).

4. For related connections between the languages of cloning and transcendence, see articles by W. J. T. Mitchell and Seaman.

5. For a survey of additional young adult science fiction featuring human cloning

and genetic engineering, see Crew's and Ostry's paired 2004 articles in *The Lion and the Unicorn*. Additional recent examples, all of which add elements of thriller and teen romance, include Dan Krokos's *False Memory* (2012), Stacey Kade's *The Rules* (2013), Jennifer Rush's *Altered* (2013), and Debra Dockter's *Deadly Design* (2015).

6. In the same period as *Cloned Lives*, Herbert's *The Eyes of Heisenberg* (1966) imagines "a world of gene-stamped sameness" (39) in which characters fear they are no longer unique and are disgusted by the prospect of impregnating a woman "in the ancient way" (80). Likewise, the future of Mitchison's *Solution Three* (1975) has lost something through "the strengthening" of cloning and genetic modification, and by its conclusion, the protagonists are ready to reintroduce "the uncertainties of meiosis" (109). Closer to the turn of the century, Michael Marshall Smith's *Spares* details a postapocalyptic society in which imprisoned clones are sexually exploited, and the narrator reflects, "We used to have religion but now we had code" (326). Often, in fact, the scientific work itself is religiously motivated: John Case's *The Genesis Code* features a devious priest-turned-scientist who quietly implants clones of Jesus Christ that zealots view as "abominations" and attempt to murder, while James BeauSeigneur's *In His Image* is a biotech-enabled rapture story featuring a messianic figure cloned from cells taken from the Shroud of Turin.

7. Wolfe's "The Fifth Head of Cerberus" (1972) is another cloning tale from this period that differentiates in some measure between cloned individuals. As the narrator's "father" observes near the story's conclusion, "You see, Doctor, your supposition that we are all truly the same individual will have to be modified. We have our little variations" (72).

8. The figurative manner in which I use the term *queer* in this section relies on the parallels between actual responses to nontraditional sexual relationships and the trilogy's fictional reactions to relationships between humans and Oankali. If one focuses on homosexuality in the novels more directly, though, there is no evading Canavan's critique: "The Oankali's vision of the future may be less sexist and less racist than our present, but it is *more* homophobic and transphobic, with a sort of compulsory heteronormativity that sees both the backward-looking reactionary male humans and their Oankali captors agreeing that all the humans are to pair off in monogamous heterosexual couples in order to have children" (*Octavia E. Butler* 107).

9. Indeed, it is worth noting how Lilith awakens other humans from podlike "suspended animation plant[s]" (127) that uncannily recall the organic clone cases featured in *Invasion of the Body Snatchers*.

10. Rachel Greenwald Smith offers a nuanced examination of the trilogy's ecological significance, recognizing how Butler's aliens reflect more elements of humanity than are initially apparent. "Oankali acquisitiveness," she emphasizes, "is not merely generative; it is also catastrophically destructive" (560). Similarly, Bruce Clarke reads the trilogy through Haraway's *Modest_Witness@Second_Millennium*, showing that the Oankali assimilation of humanity "enacts the cultural repercussions of a technologized bioscience that has shredded any notion of inviolable natural or cultural realities withstanding its transformative doctrines and techniques" (164).

11. For more on the trilogy's treatments of sexuality, disability, and race, see Obourn's insightful assessment. For an excellent treatment of queerness and cloning in SF cinema, see Turner's argument that cloning scenarios reimagine the maternal and the reproductive, both reflecting and questioning "the all-too-familiar images that naturalize women and their bodies as infantile, mad, wild, and animal-like" (112). And for a broad survey of queerness and cloning in print science fiction, see Wasson's study of not just the "surprisingly rare" story of queer clone love, but also heterosexual clone encounters, which are closely related because of their interest in "love-for-the-self" (Wasson, "Love" 133).

12. Like earlier critics, Nanda notes how Butler's character diverges from a traditional, "idealized" Eve toward the rabbinic character of Lilith, whom Adam instructs to "lie below him" (779).

13. The closest Butler comes to genetic realism is the 1987 short story published as "The Evening and the Morning and the Night." Canavan's "Life without Hope?" considers its relationship to Huntington's disease, as does one of my articles, "Determined Agency."

14. Canavan's *Octavia E. Butler* powerfully challenges the trend in Butler criticism of glossing over Oankali abuses of other species. The unhappy conclusion in which he joins Joan Slonczewski is that "the Oankali, for all their supposed difference and for all their self-inflating rhetoric, are as much the final perfection of human destructivity as an alternative to it" (Canavan 119).

15. In a 2000 interview on "Persistence" for the SF magazine *Locus*, Butler recalled, "I used to despise religion. I have not become religious, but I think I've become more understanding of religion. And I'm glad I was raised as a Baptist, because I got my conscience installed early. I've been around people who don't have one, and they're damned scary. And I think a lot of them are out there running major corporations! How can you do some of the things these people do if you have a conscience? So I think it might be better if there were a little *more* religion, in that sense" (emphasis in original).

16. Few commentaries on the trilogy consider Akin's status as a modified Christ figure, but Judith Lee observes that Akin mirrors Christ's "dual divine-human nature" (175).

17. The trilogy is sometimes so harsh in referencing biblical narrative that Butler can seem virulently antireligious. It is no coincidence that the man with whom the Oankali first pair Lilith is named after two New Testament figures, especially when we consider Butler's comment that "I like to use names that work with who my characters are" (Wolmark 32n13). Paul Titus warns Lilith about other men who "will want to be cavemen—drag you around, put you in a harem, beat the shit out of you" (93) and then attempts to rape her himself.

18. Clayton offers a similarly complex reading of *Xenogenesis*'s openness to genetic engineering alongside a critique of sexist exploitation, arguing that the trilogy "supports genetic manipulation of the species but does not hide its relationship to other kinds of violence against women" ("Ridicule" 329).

19. Theologically, synergism can be understood in contrast with monergism, which emphasizes "irresistible grace," or the divine role in salvation over against any individual human response. Synergism's more cooperative, two-way vision is closely associated with Arminian theology and is especially influential in Lutheran, Methodist, Reformed, Catholic, and Eastern Orthodox churches.

20. Belk highlights one of the most intense critiques of Christianity in Butler's trilogy—namely, its cultivation of profligate consumerism. Pointing to an *Adulthood Rites* passage in which Akin "tastes" a holographic image of Jesus made of plastic and immediately rejects it, finding "more poison packed together in one place than I've ever known" (388), Belk makes a compelling argument that the "erotic" in Butler encompasses not only the sexual but also a bodily intuition that exceeds logic.

21. Those interested in the origins of *Xenogenesis* and Butler's prescient views of sociobiology should see Johns's study of her extensive reliance on the work of E. O. Wilson. Relatedly, Bruce Clarke builds on Haraway's description of the Oankali as "extraterrestrial lovers/rescuers/destroyers/genetic engineers" (179) as well as Slonczewski's assessment that "unlike the vast majority of alien abduction tales, *Dawn* actually presents a biologically plausible explanation for why the Oankali need to interbreed with humans." As if anticipating Morton's work on hyperobjects, Clarke concludes that "the master key to what Butler's narrative has accomplished, then, is the way that she channels all that macrocosmic extraterrestrial transcendentalism through the microcosm, 'a rush of individual cells'" (192).

22. This minimalist aesthetic approach garners *Moon* an assessment by Springer as "hard science fiction" deserving comparison to *2001: A Space Odyssey* and *Blade Runner*, but her article explains only that "*Moon*'s inclusion of cloning is based on previous scientific achievements, such as the successful cloning of the sheep Dolly" (41).

Chapter 3

1. Additional early examples of genetic realism include Simon Mawer's *Mendel's Dwarf* (1998) and (obliquely) the title story in Andrea Barrett's *Ship Fever: Stories* (1996). I especially recommend Byers's *Long for This World* (2003), a gripping but sober tale of a physician who uses an unapproved gene therapy to try to save a young man from a disease resembling progeria. The novel recognizes that "the study of how proteins folded and unfolded and why they worked and stopped working would occupy geneticists for decades into the future—that was the real work that would be waiting for them all after the fanfare of the genome project had faded" (130).

2. Audiences sometimes wonder if the show's setting is to be understood as some sort of generic North American no-place. Showrunner Graeme Manson shared with me that this was never the creators' intention and that while the show initially avoided explicitly Canadian visual references, recognizable Toronto structures were eventually allowed to appear in the mise-en-scène.

3. Here my approach dovetails with that of a forthcoming book chapter by Vint, "Commodifying Life," in which she shows how *Orphan Black* aligns Ethan Duncan with Wells's Moreau and Henrik Johanssen with Shelley's Frankenstein.

4. Smith's *White Teeth* also utilizes the term *abomination* to describe an object of misunderstanding and fear, the FutureMouse created by Marcus Chalfen to live an exactly prescribed life span. Millat Iqbal calls it an "abomination," while his brother Magid sees it as an effort toward "correcting the Creator's mistakes" (383).

5. If my interpretation of the shower scene seems excessively Freudian, consider Lepore's observation that "the show's go-to wound is the puncture: the act of penetration." Related moments include Rachel's stabbing of her mother and then Sarah at the end of season four.

6. See Jer. 31:15 and its citation by Matt. 2:18, both of which describe moments of lamentation in Israel's history (first the Babylonian exile, then a holocaust occasioned by Herod's fear of a Jewish messiah).

7. This passage appears to grow even more significant in foreshadowing the creature introduced in the show's fifth season. That narrative was still unfolding during this book's final editing process; see my series of online articles about season five in the *Los Angeles Review of Books*.

8. Herter explained to Newton that part of her job is keeping the show's treatment of biotechnology and bioethics in the interrogative mode: "You won't have to fictionalize things if you ask these kinds of questions. You'll be fictionalizing if you try to answer them." Recognizing the difference between science as methodology and as metaphysic, Herter resists any Gouldian expectation that science and religion must be separated: "That they are somehow two different magisterial domains that can't cross each other is just so fundamentally untrue."

9. As Thurtle demonstrates in *The Emergence of Genetic Rationality*, the work of mapping inheritance was once entirely a narrative art, but that changed with Mendel and a new genetics defined by a chart-oriented recordkeeping. On the history of Cold Spring Harbor Laboratory and the Eugenic Records Office, see also Kevles.

10. As Ishiguro explained to Wong and Crummett, he began writing the novel in 1990 as a tale about students living in the countryside, separated from any university, who become sick through an encounter with some kind of nuclear device. The novel was always about the transience of a human life, but Ishiguro set it aside until 2001, when talk about Dolly the sheep was in the air and he began imagining the students as clones. As Ishiguro said, "I could see a metaphor here. I was looking for a situation to talk about the whole aging process, but in such an odd way that we'd have to look at it all in a new way" (213).

11. Some interpreters of Ishiguro's novel have approached it as a more literal attempt to imagine a future in which it is commonplace to use human clones as unquestioning organ donors. Wasson argues that this "haunting novel . . . helps us to imagine the way covert lifelong pressure in such a relationship [between a presumptive organ donor and society] could shape a mind into unwilling

surrender" ("Butcher's Shop" 80), and Ivan Stacy laments that "the clones fail to bear witness to their own condition" (238). Griffin, however, shares my sense that "[the novel's] function is not to actualize science in quasi-mimetic fashion but to comment critically on the history of the present" (653), an orientation that she goes on to align with the work of Jameson and Haraway. Query states simply that *Never Let Me Go* is "far less a novel about what we do not know or what might be, a speculative fiction, than about our acceptance of what we already know and are" (161).

12. Other varieties of genetic fantasy turn this paradigm on its head. Nancy Kress's *Beggars in Spain* (1993), *Beggars Ride* (1994), and *Beggars and Choosers* (1996) are just as concerned with social inequality and questions of ableism as Ishiguro's novel, but in her storyworld, the genetically engineered children seem the fortunate ones.

13. Eatough makes the ironic point that the existence of this "possible," coupled with Ruth's genetic determinism, allows her to "devalue her own body," in part because she "refuse[s] to clearly differentiate between [it] and other bodies" (149). Gill uses this hunger for identification to comment on the novel's treatment of racial difference, observing that in "searching for recognition in the face of a genetically similar other, Kathy instead finds the recognition she has been seeking in the facial expression of the novel's only racially differentiated character, Miss Emily's Nigerian carer George" (847).

14. For all of the absurdity of the children's mythology, the real process of DNA replication requires that the double helix be "unzipped" so that one of the strands can be copied. Like many things these young clones know without quite understanding, their metaphor is more appropriate than it sounds.

15. Haran et al. provide fascinating evidence of this slippage. Demonstrating "temporal contraction" in popular perceptions of very diverse biotechnologies, their research participants "spoke of cloning being responsible for hand transplants and growing bladders and other forms of tissue engineering such as that associated with the widely circulated image of a human ear growing on the back of a mouse. Such comments illustrate and register the condensation of the timeframe for medical innovations associated with cloning and the general sense that cures deriving from therapeutic cloning are not only imminent, but have already been realised" (48).

16. The novel's setting is especially worth remembering now that Ishiguro has expressed public alarm over the possibilities enabled by CRISPR. Late in 2016, he told Hannah Devlin, "We're coming close to the point where we can, objectively in some sense, create people who are superior to others." However well intended, this activism could make it even easier to forget that *Never Let Me Go* is set in England's *past*; in fact, Devlin's piece itself misleadingly summarizes the novel as taking place in "a dark future." The novel's real power lies not in a literal challenge to some future fantasy of exploiting human clones but in its indictment of enduring patterns of actual scapegoating.

17. A rare critical treatment of Jones's novel, Vint's article in *The New York Review of Science Fiction* examines institutional barriers to women's advancement in scientific careers.

18. *White Teeth*'s and *Middlesex*'s interests in immigration, scientism, and religious cults make them further illustrations of Justin Neuman's argument that "some of the most trenchant and far-reaching critiques of secularist ideologies, as well as the most exciting and rigorous inquiries into the legacies of the religious imagination, take place where we might least expect them: in the pages of contemporary novels composed by a transnational group of writers commonly identified as non- or even antireligious" (xi). At the same time, as Chu argues, early twenty-first-century novels of genetic realism like *Middlesex* might also suggest a turning point for the ethnic novel, wherein biological inheritance is as significant as "reproduction of culture and community or the recovery of history" (208).

19. See, for example, readings of *White Teeth* by Trimm and Watts, and of *Middlesex* by Hsu; Lee; and Shostak.

20. I refer to *Middlesex*'s narrator as Cal whenever the events in question take place after his gender reidentification or when we

are discussing him as narrator (since the narration takes place in a later time period). When discussing the main character's thoughts or actions before Callie becomes Cal, however, I refer to her as Callie.

21. Smith uses the same technique to preview Ryan Topps, the moped rider whose crash results in Clara Bowden's loss of her top row of teeth and her eventual marriage to Archie Jones; Mo/Mohammad Hussein-Ishmael, the halal butcher who eventually joins KEVIN; and of course Dr. Perret, the Nazi eugenicist, introduced much earlier as Dr. Sick.

22. McMann's reading of Smith's novel takes its interest in eugenics and genetics far more seriously than most.

23. John Money was a psychologist at Johns Hopkins who became particularly infamous for his treatment of David Reimer. After the twenty-two-month-old boy's circumcision was horribly botched, Money convinced Reimer's parents to castrate him, ignore his chromosomal sex, and raise him as a female. Reimer went public as an adult with his cautionary tale and later committed suicide.

24. Another young woman in the novel who strenuously objects to biological determinism is Archie's daughter, Irie. She is "unwilling to settle for genetic fate" (222) and so "intent upon fighting her genes" (227) that in order to attract Millat Iqbal, she decides she must have straight hair. The ironic result is that a young British woman attempts to woo the son of a Bengali Muslim from Bangladesh with hair taken from an Indian immigrant—and the hair immediately begins falling out, as if the graft simply would not take.

25. As if extending the logic of epigenetics' stress on environmental context, Eugenides's narrator also remarks, "Parents are supposed to pass down physical traits to their children, but it's my belief that all sorts of other things get passed down, too: motifs, scenarios, even fates" (109).

Chapter 4

1. The term *metafiction* was coined in 1970, though of course its applications extend much further into literary history. Birgit Neuman and Nünning provide a concise overview of its ongoing development.

2. Book two of the five-book "Deluxe" hardbound omnibus edition, page 130. This format also applies to all additional citations of *Y: The Last Man*. Since boldfaced/italicized text in speech balloons is so common in comics, I do not append the label "emphasis in original" each time.

3. Brown provides an extensive treatment of LGBTQ issues in *Y: The Last Man*, especially in "Safeword" and "Girl on Girl."

4. This is likely an allusion to Gordon Rattray Taylor's *The Biological Time Bomb* (1968), an early work of popular science (some would say "pseudoscience") that might have also inspired Le Guin's "Nine Lives." A similar premise appears in Gwyneth Jones's *Life*.

5. Studies of asexual organisms often demonstrate how sexual fertilization's unpredictability provides species a valuably diverse gene pool in warding off new environmental threats.

6. As more works of genetic metafiction appear, it might be worth thinking especially about how they reframe relationships between image and word. Other comics engaging genetics heavily include Jonathan Hickman and J. M. Ringuet's *Transhuman* (2008), Richard Marazano and Jean-Michel Ponzio's *Genetiks™* (2012), and Greg Rucka, Michael Lark, and Santi Arcas's *Lazarus* (2013). None of these, however, is as self-reflexive as *Y: The Last Man* or Steve Tomasula's *VAS: An Opera in Flatland* (2002), an experimental work of new media that intersperses narrative with a twenty-five-page list of nucleotides found on chromosome twelve, ties between early-twentieth-century Fitter Family Contests and contemporary beauty pageants, and juxtaposed advertisements for DNA testing, egg and sperm auctions, and cosmetic surgery.

7. For those fascinated by media's influence on narrative structure, Millington makes a compelling argument that the shift from "24-page floppies" to omnibus collections (like those in which *Y: The Last Man* is now most easily available) eliminates significant aspects of the reading experience.

8. For pop culture references in *Y: The Last Man*, see Buesing.

9. In reviewing *The Year of the Flood*, Jameson proposed another organizational schema that might be expanded to cover the whole trilogy. Focusing on the technocapitalist near-future society Atwood imagines, Jameson explained, "*Oryx* gave us the view of this system from the inside and as it were from above," while "*The Year of the Flood* gives us the view from below." Perhaps *MaddAddam*, then, provides the view from *afterward*? Those interested in Atwood's structural innovations should also see Snyder's "'Time to Go,'" which explains how a postapocalyptic setting allows for "fundamentally unwitnessable" (479) fantasies and greater awareness of "retrodetermination" (485).

10. Unfortunately, the US publisher feared it sounded like a "hippy New Age cult."

11. See also Herbert's *The White Plague* (1982), in which a main character avenges an IRA bombing that killed his family by loosing a global pandemic.

12. Barzilai provides valuable context for understanding Crake's "revenge tragedy" by demonstrating how deeply Atwood engages Shakespeare's *Hamlet* in constructing this character.

13. See McHugh for an excellent treatment of the *MaddAddam* trilogy's concern with food production and access.

14. For those curious about transgenic organisms and the swiftly emerging field of biocommerce, see Waldby and Mitchell's *Tissue Economies* as well as other work by Mitchell, Burgess, and Thurtle.

15. My response to Idema's article "Toward a Minor Science Fiction," which foregrounds Atwood's trilogy, is similar. Citing only *Oryx and Crake*, Idema includes Atwood with Shelley and Wells as "mostly concerned with warning against the possible dangers of technoscience, leaving alternative, positive uses relatively untapped" (52). I agree that Atwood is more cautionary than enthusiastic about biotech, but the full trilogy makes her approach more nuanced.

16. Adam One is not the only character that readers must reevaluate through the perspectives provided by *MaddAddam*. For example, Jimmy's inability to take seriously the young women he seduced in his former life might receive little overt criticism in *Oryx and Crake*, but *MaddAddam* reveals the sources of the whisperings from "some tart he once bought" (*Oryx and Crake* 11). She is Ren, his former girlfriend, whom he'd long since forgotten.

17. I am not alone in arguing that Atwood is far more serious and sympathetic in imagining Adam One and the God's Gardeners than many assume. Assessing *The Year of the Flood*, Bergthaller notes, "At the outset of the narrative, Adam One's abrupt transitions from the biological to the religious register appear mostly as comedic; yet as the narrative progresses, the reader also comes to appreciate the combination of humility, generosity, and hardheaded, practical shrewdness which they reflect. Especially towards the end of the novel, the clownish aspects of his character give way to something approaching genuine saintliness" (740). Hoogheem also interprets Atwood's second novel as genuinely envisioning an ecologically attuned theism; there are serious tensions and much humor inherent in the God's Gardeners' cosmology, which "lurch[es] between a rhapsodic panentheism (or at least a very deep sacramentality) and an ascetic dualism," but at the same time, "the Gardeners' kitchen-sink approach to orthodoxy serves an unwavering dedication to orthopraxis" (62).

18. Just as the "Adam" found in Genesis is both a proper noun and representative of an entire species ("*ad-dam*" is Hebrew for "humanity"), Atwood's puppeteer in *MaddAddam* might stand for contemporary humanity and its capacity for madness.

19. Chapter 2 notes parallels between Atwood's trilogy and Kate Wilhelm's *Where Late the Sweet Birds Sang*, but the complexity of Blackbeard's character is one of many places that the gap is visible. Atwood zooms in far closer to her posthuman characters.

20. Excerpts from these performances are featured in the documentary *In the Wake of the Flood* (2010).

21. The term *fundamentalist* was first used widely in the wake of an early

twentieth-century pamphlet, *The Fundamentals*. Though now applied more broadly, it references a specific lineage of US Protestantism that has developed an increasingly literalistic biblical hermeneutic and that regularly opposes itself to "secular" American culture.

Chapter 5

1. Atwood's 2015 contribution to the Future Library in Oslo, *Scribbler Moon*, was joined in 2016 by David Mitchell's *Me Flows What You Call Time*. One text will be added each year until the entire trove is printed in 2114.

2. This distinction between scientific curiosity and obsession with control is no secret among Powers scholars. Articles by Hermanson ("Chaos"), Hume, and Strecker all cite at least one of the passages in *The Gold Bug Variations* that I have highlighted.

3. The foundational manifestoes of this movement remain provocative. See especially Harding and Merchant.

4. This chapter concentrates on moments in which science edges toward its own brand of religious transcendence, but that should not suggest that Powers ignores traditional religious manifestations. Near the end of *The Gold Bug Variations*, genetic research and belief in young-earth creationism strike discordant tones; in *The Echo Maker*, Gerald associates religion with "childish things" (309). *Generosity* includes Powers's most direct assault on religious anti-intellectualism when Mike Burns, an evangelical megachurch minister, claims that evolution is a lie and that it cannot be "just chance" that "this recipient of God's unstoppable love just *happens to be* Christian" (207, emphasis in original). This is the worst sort of appropriation, as Thassa is broadly tolerant of religion, but figures that "if there is a God, he is just laughing at every religion we invent!" (48).

5. There has been strikingly little movement in Gallup's polling on this issue over the last three decades. As of 2014, 42 percent of respondents affirmed that "God created humans pretty much in their present form at one time within the last 10,000 years or so." The number was 44 percent in 1982.

6. Powers's work is deeply concerned with a question asked by Broglio (following Agamben, following Heidegger): "What then would man and animals be who are outside of history and outside of the anthropological machine which separates the human from his animal nature?" (3).

7. This chapter approaches "creative nonfiction" from the fictional side of the relationship, only minimally engaging the growing body of work called genetic life-writing. For overviews of that genre, see Couser's and O'Riordan's pieces in *Biography* and Jurecic's concise "The Art of Medicine." For representative examples of the genre itself, see work by Angrist, Frank, Pinker, and Wexler.

8. Dewey suggests that this simultaneity has a distinct effect: "Unlike postmodern narratives that taunt the reader by reminding us ad-nauseum that storytelling is necessarily defeated by closure, here Powers invites us to keep the story alive" (224).

9. As early as the novel's second page, the narrator observes of the commuters around Russell: "Just brushing against them in memory makes me panic." A few paragraphs later, we find what on rereading becomes a confession: "I know this story, like I wrote it myself" (4). Other metanarratival moments across the novel become meaningful only once we understand the narrator's identity—a protest that "someone has invented the scene just to create rising action" (104), a fear that "the story wanted to break away, lose itself, escape altogether its birthright plot" (269).

10. Powers pointed to the influences of Hardy and Joyce in a 2003 interview with Fuller, wherein he asked, "Why not see if realism and metafiction can't leverage one another to create even more possibilities? Why not?" (113).

11. Kurton's objectivism is closely related to the demands of Tonia's producer to "interview" Thassa, not "befriend" her (234). It also dovetails with Candace's surrender to colleagues' expectations and her attempts to hide behind a counselor mask when Thassa needs her most, even as

Russell recognizes her rhetoric's "pop-psych Serenity Prayer" (263).

12. Kurton's theory of fiction echoes Darwinian literary criticism, in which "evocritics" argue that human universals can provide a corrective for overattentiveness to geographical and temporal context. Among their principal claims, as expressed in Boyd, is that evolution selected for fiction because "it has sharpened social cognition and extended our capacity to think beyond the here and now, to entertain other possibilities and not simply accept the given" (206). This makes some anthropological sense, but it risks mobilizing an all-too-familiar secularization narrative whereby religion becomes the most sentimental of fictions, a childish fantasy to be surrendered as humanity matures. This is almost exactly Kurton's line. "For most of human history," he announces, "when existence was too short and bleak to mean anything, we needed stories to compensate" (140).

13. Schaefer acknowledges that Russell could be the narrator "possibly even of the entire novel" (280) but does not further investigate the implications of that rereading.

14. Writing for *The Hastings Center Report*, Lantos rightly observes how the text depicts "science as the ultimate redemption story of our age." However, Lantos misinterprets Powers as affirming Kurton's brand of scientism. He assumes the novel supports Kurton's view of contemporary novels as "obsolete," valuable "merely as inefficient mood-stabilizing devices." As a result, when Kurton debates a novelist and then a bioethicist, Lantos decides, "Powers clearly prefers the witty, visionary scientist to the dowdy bioethicist" and that "ethics, for Powers, is old and tired. Science is young, new, and fresh." By contrast, I read *Generosity* as using Kurton to expose the danger of genetics becoming a religion unto itself.

15. Those who know Powers's broader oeuvre might hear autobiographical elements in Els's confession. In a 2008 interview with Burn, Powers looked back on his own career and explained how William Gaddis was "the first to make me realize that a large-scale work of fiction could be both highly, schematically constructed and deeply, passionately inhabited" (173).

16. After the publication of *Orfeo*, when asked by Jackson how soon he might turn up on Twitter, Powers protested, "I'm a long-form guy." However, it appears that either Powers or his publisher set up an actual Twitter account to broadcast Els's messages during the five months before the novel's appearance. At https://twitter.com/TerrorChord, twenty-two messages almost exactly match those found in the novel's first eighty-nine pages. The account is listed as following, among others, Salman Rushdie, Joyce Carol Oates, William Gibson, and yes, Margaret Atwood. The tweets are unsigned, making no reference to *Orfeo* or its author.

Coda

1. In describing the heptapods' nonlinear orthography, *Arrival* directly cites the Sapir-Whorfian hypothesis, which in a strong form suggests that the particulars of language determine how one thinks. As with my critique of genetic determinism, I would suggest that the linguistic variation overextends a valuable awareness that the particulars of language really do shape experience.

2. Chiang's "Story of Your Life" further explains Fermat's principle, the physical paradox whereby a ray of light originating from land reaches a point offshore and underwater fastest via not a direct path but a bent one (since light travels more rapidly through air than water). The story relies on the distinction between causal, linear, sequential thinking, which focuses on explaining the refraction of the light, and teleological, holistic, simultaneous thinking, which imbues purpose to the light: "The ray of light has to know where it will ultimately end up before it can choose the direction to begin moving in" (156).

3. *Arrival* participates in a long tradition of SF first-contact stories that invite theological interpretations, many of which

also foreground parent-child relationships. Literary examples include H. G. Wells's *The War of the Worlds* (1898), Arthur C. Clarke's *Childhood's End*, Mary Doria Russell's *The Sparrow* (1996), and Liu Cixin's *The Three-Body Problem* (English translation, 2014) and its sequels. In cinema, think of *2001: A Space Odyssey*, *Close Encounters of the Third Kind* (1977), *Contact*, and *Interstellar* (2014).

4. For the story of Hannah's pregnancy, see 1 Sam. 1; for the story in which Saul convinces the witch of Endor to help him consult the ghost of Samuel, see 1 Sam. 28.

Works Cited

Aaron, Jason, et al. *X-Men: Schism* (*X-Men: Schism* #1–5 and *X-Men: Regenesis* #1). Marvel, 2011. Print.

Alaimo, Stacy. *Bodily Natures: Science, Environment, and the Material Self*. Indiana UP, 2010. Print.

Angrist, Misha. *Here Is a Human Being: At the Dawn of Personal Genomics*. HarperCollins, 2010. Print.

Arrival. Dir. Denis Villeneuve. Perf. Amy Adams, Jeremy Renner, Forest Whitaker. 21 Laps Entertainment, 2016.

Atwood, Margaret. *The Blind Assassin*. Anchor, 2000. Print.

———. *The Handmaid's Tale*. 1985. Anchor, 1998. Print.

———. "In the Heart of the Heartland." Rev. of *The Echo Maker*, by Richard Powers. *New York Review of Books*. 21 Dec. 2006. Print.

———. *MaddAddam*. Random House, 2013. Print.

———. *Oryx and Crake*. Doubleday, 2003. Print.

———. *The Tent*. Anchor, 2006. Print.

———. *The Year of the Flood*. Anchor, 2009. Print.

Barnes, Barry, and John Dupré. *Genomes and What to Make of Them*. U of Chicago P, 2008. Print.

Barrett, Andrea. *Ship Fever: Stories*. Norton, 1996. Print.

Barzilai, Shuli. "'Tell My Story': Remembrance and Revenge in Atwood's *Oryx and Crake* and Shakespeare's *Hamlet*." *Critique* 50.1 (2008): 87–110. Print.

BeauSeigneur, James. *In His Image*. 1988. Warner, 2003. Print.

Belk, Nolan. "The Certainty of the Flesh: Octavia Butler's Use of the Erotic in the *Xenogenesis* Trilogy." *Utopian Studies* 19.3 (2008): 369–89. Print.

Bergthaller, Hannes. "Housebreaking the Human Animal: Humanism and the Problem of Sustainability in Margaret Atwood's *Oryx and Crake* and *Year of the Flood*." *English Studies* 91.7 (2010): 728–43. Print.

Bérubé, Michael. *Life as Jamie Knows It: An Exceptional Child Grows Up*. Beacon, 2016. Print.

———. *Life as We Know It: A Father, a Family, and an Exceptional Child*. Vintage, 1996. Print.

Blackford, Russell. *Humanity Enhanced: Genetic Choice and the Challenge for Liberal Democracies*. MIT P, 2014. Print.

Blade Runner. Dir. Ridley Scott. Perf. Harrison Ford, Rutger Hauer, Sean Young. Warner Bros., 1982.

Boyd, Brian. *On the Origin of Stories: Evolution, Cognition, and Fiction*. Harvard UP, 2009. Print.

Branch, Lori. "Postsecular Studies." *Routledge Companion to Literature and Religion*. Ed. Mark Knight. Routledge, 2016. 91–101. Print.

Broglio, Ron. "When Animals and Technology Are beyond Human Grasping." *Angelaki* 18.1 (2013): 1–9. Print.

Brooks, Peter. *Reading for the Plot: Design and Intention in Narrative*. Harvard UP, 1984. Print.

Brown, Lyndsay. "Yorick, Don't Be a Hero: Productive Motion in *Y: The Last Man*." *ImageTexT* 3.1 (2006). Web. <http://www.english.ufl.edu/imagetext/archives/v3_1/brown/>.

Buesing, Dave. "Every Single Pop Culture Reference in *Y: The Last Man*." *Comic Book Herald*. 17 Sept. 2013. Web. <http://www.comicbookherald.com/every-single-pop-culture-reference-in-y-the-last-man/>.

Burn, Stephen J. "An Interview with Richard Powers." *Contemporary Literature* 49.2 (2008): 163–79. Print.

Butler, Octavia E. *Lilith's Brood* (*Xenogenesis*: *Dawn*, *Imago*, and *Adulthood Rites*, 1987–89). Grand Central, 2000. Print.

———. "Octavia E. Butler: Persistence." *Locus* 44.6 (2000): 4, 75–78. Print.

Byers, Michael. *Long for This World*. Houghton Mifflin, 2003. Print.

Canavan, Gerry. "Hope, but Not for Us: Ecological Science Fiction and the End of the World in Margaret Atwood's *Oryx and Crake* and *The Year of the Flood*." *LIT* 23 (2012): 138–59. Print.

———. "Life without Hope? Huntington's Disease and Genetic Futurity." *Disability in Science Fiction: Representations of Technology as Cure*. Ed. Kathryn Allan. Palgrave Macmillan, 2013. 169–87. Print.

———. *Octavia E. Butler*. Illinois UP, 2016. Print.

Case, John. *The Genesis Code*. Fawcett, 1997. Print.

Cherryh, C. J. *Cyteen*. Warner, 1988. Print.

———. *Downbelow Station*. DAW, 1981. Print.

———. *Regenesis*. DAW, 2009. Print.

Chiang, Ted. "Story of Your Life." *Stories of Your Life and Others*. Tor, 2002. 117–78. Print.

Chu, Patricia E. "The American Genome Project: A Biopolitical History of the Contemporary Ethno-Racial Novel." *American Literary History* 25.1 (2013): 205–16. Print.

Church, George, and Ed Regis. *Regenesis: How Synthetic Biology Will Reinvent Nature and Ourselves*. Basic, 2012. Print.

Cixin, Liu. *The Three-Body Problem*. 2006. Trans. Ken Liu. Tor, 2014. Print.

Clarke, Arthur C. *Childhood's End*. 1953. Del Rey, 1987. Print.

Clarke, Bruce. *Posthuman Metamorphosis: Narrative and Systems*. Fordham UP, 2008. Print.

Clayton, Jay. *Charles Dickens in Cyberspace: The Afterlife of the Nineteenth Century in Postmodern Culture*. Oxford UP, 2003. Print.

———. "Genome Time." *Time and the Literary*. Ed. Jay Clayton, Karen Newman, and Marianne Hirsch. Routledge, 2002. 31–59. Print.

———. "The Ridicule of Time: Science Fiction, Bioethics, and the Posthuman." *American Literary History* 25.2 (2013): 317–40. Print.

Close Encounters of the Third Kind. Dir. Steven Spielberg. Perf. Richard Dreyfuss, François Truffaut, Teri Garr. Columbia, 1977.

Code 46. Dir. Michael Winterbottom. Perf. Tim Robbins, Samantha Morton. BBC, Kailash, and Revolution, 2003.

Collins, Suzanne. *The Hunger Games*. Scholastic, 2008. Print.

Comfort, Nathaniel. *The Science of Human Perfection: How Genes Became the Heart of American Medicine*. Yale UP, 2012. Print.

Contact. Dir. Robert Zemeckis. Perf. Jodie Foster, Matthew McConaughey, Tom Skerritt. Warner Bros., 1997.

Couser, G. Thomas. "Genome and Genre: DNA and Life Writing." *Biography* 24.1 (2001): 185–96. Print.

Coviello, Peter, and Jared Hickman. "Introduction: After the Postsecular." *American Literature* 86.4 (2014): 645–54. Print.

Crew, Hilary. "Not So Brave a World: The Representation of Human Cloning in Science Fiction for Young Adults." *The Lion and the Unicorn* 28.2 (2004): 203–21. Print.

Csicsery-Ronay, Istvan, Jr. *The Seven Beauties of Science Fiction*. Wesleyan UP, 2008. Print.

"Darwin on Stage and Screen." *BBC: Newsnight Review*. 11 Sept. 2009. Print.

Derrida, Jacques. *Acts of Religion*. Routledge, 2002. Print.

Devlin, Hannah. "Kazuo Ishiguro: 'We're Coming Close to the Point Where We Can Create People Who Are Superior to Others.'" *The Guardian*. 2 Dec. 2016. Print.

Dewey, Joseph. "Dwelling in Possibility: The Fiction of Richard Powers." 1996. *Twayne Companion to Contemporary Literature in English*. Ed. R. H. W. Dillard and Amanda Cockrell. Vol. 2. Twayne, 2002. 213–25. Print.

Dickenson, Donna. *Me Medicine vs. We Medicine: Reclaiming Biotechnology for the Common Good*. Columbia UP, 2013. Print.

District 9. Dir. Neill Blomkamp. Perf. Sharlto Copley, David James, Jason Cope. TriStar, Block/Hanson, and WingNut, 2009.

Dockter, Debra. *Deadly Design*. G. P. Putnam's Sons, 2015. Print.

Eatough, Matthew. "The Time That Remains: Organ Donation, Temporal Duration, and Bildung in Kazuo Ishiguro's *Never Let Me Go*." *Literature and Medicine* 29.1 (2011): 132–60. Print.

Elysium. Dir. Neill Blomkamp. Perf. Matt Damon, Jodie Foster, Sharlto Copley. TriStar, MRC, and QED International, 2013.

Erlich, Richard. "Le Guin and God: Quarreling with the One, Critiquing Pure Reason." *Extrapolation* 47.3 (2006): 351–79. Print.

Eugenides, Jeffrey. *Middlesex*. Picador, 2002. Print.

"Eve." *X-Files*. Ten Thirteen and Fox, North Vancouver. 10 Dec. 1993.

"Evolution, Creationism, Intelligent Design." *Gallup*. May 2014. Print.

"Faint Hope." *ReGenesis*. Shaftsbury Films, Toronto. 28 Nov. 2004.

Farmer, Nancy. *The House of the Scorpion*. Athenaeum, 2002. Print.

Fawaz, Ramzi. "'Where No X-Man Has Gone Before!' Mutant Superheroes and the Cultural Politics of Popular Fantasy in Postwar America." *American Literature* 83.2 (2011): 355–88. Print.

Fawcett, John, and Graeme Manson, creators. *Orphan Black*. Temple Street Productions and BBC America, 2013–17.

Ferreira, Maria Aline Salgueiro Seabra. *I Am the Other: Literary Negotiations of Human Cloning*. Praeger, 2005. Print.

Fessenden, Tracy. "The Problem of the Postsecular." *American Literary History* 26.1 (2014): 154–67. Print.

Feynman, Richard P. "There's Plenty of Room at the Bottom: An Invitation to Enter a New Field of Physics." *Caltech Engineering and Science* 23.5 (1960): 22–36. Print.

Foer, Jonathan Safran. "Jeffrey Eugenides." *Bomb* 81 (Fall 2002): 74–80. Print.

Frank, Lone. *My Beautiful Genome: Exposing Our Genetic Future, One Quirk at a Time*. 2010. Trans. Russell Dee. Oneworld, 2011. Print.

Fuller, Randall. "An Interview with Richard Powers." *Missouri Review* 26.1 (2003): 97–113. Print.

Gagnier, Regenia. "Freedom, Determinism, and Hope in Little Dorrit: A Literary Anthropology." *Partial Answers* 9.2 (2011): 331–46. Print.

Gattaca. Dir. Andrew Niccol. Perf. Ethan Hawke, Uma Thurman, Jude Law. Columbia Pictures and Jersey Films, 1997.

"Gay Gene Isolated, Ostracized." *The Onion* 44.26. 9 April 1997. Web. <http://www.theonion.com/article/gay-gene-isolated-ostracized-4287>.

"George Church." *The Colbert Report*. 4 Oct. 2012. Web. <http://www.cc.com/video-clips/fkt99i/the-colbert-report-george-church>.

Gill, Josie. "Written on the Face: Race and Expression in Kazuo Ishiguro's *Never Let Me Go*." *Modern Fiction Studies* 60.4 (2014): 844–62. Print.

Girard, René. *The Scapegoat*. Johns Hopkins UP, 1986. Print.

Gleick, James. "When They Came from Another World." Rev. of *Arrival*, dir. by Denis Villeneuve, and *Stories of Your Life and Others*, by Ted Chiang. *New York Review of Books*. 19 Jan. 2017. Print.

Green, Ronald M. *Babies by Design: The Ethics of Genetic Choice*. Yale UP, 2007. Print.

Griffin, Gabriele. "Science and the Cultural Imaginary: The Case of Kazuo Ishiguro's *Never Let Me Go*." *Textual Practice* 23.4 (2009): 645–63. Print.

Hamer, Dean. *The God Gene: How Faith Is Hardwired into Our Genes*. 2004. Anchor, 2005. Print.

Hamner, Everett. "Determined Agency: A Postsecular Proposal for Religion and Literature—and Science." *Religion and Literature* 41.3 (2009): 24–31. Print.

———. "Epigenetic television: The Penetrating Love of 'Orphan Black.'" *Los Angeles Review of Books*. 9 June 2017. Web. <https://lareviewofbooks.org/article/epigenetic-television-the-penetrating-love-of-orphan-black/>.

Haran, Joan, et al. *Human Cloning in the Media: From Science Fiction to Science Practice*. Routledge, 2008. Print.

Haraway, Donna. "A Cyborg Manifesto: Science, Technology, and Socialist-Feminism in the Late Twentieth Century." *Simians, Cyborgs, and Women: The Reinvention of Nature*. Routledge, 1991. 149–81. Print.

———. *Modest_Witness@Second_Millennium.FemaleMan©_Meets_OncoMouse™: Feminism and Technoscience*. Routledge, 1997. Print.

———. *When Species Meet*. Minnesota UP, 2008. Print.

Harding, Sandra. *Whose Science? Whose Knowledge? Thinking from Women's Lives*. Cornell UP, 1991. Print.

Hayles, N. Katherine. "Desiring Agency: Limiting Metaphors and Enabling Constraints in Dawkins and Deleuze/Guattari." *SubStance* 30.1–2 (2001): 144–59. Print.

Heller, Peter. *The Dog Stars*. Knopf, 2012. Print.

Herbert, Frank. *The Eyes of Heisenberg*. Berkley, 1966. Print.

———. *The White Plague*. G. P. Putnam's Sons, 1982. Print.

Hermanson, Scott. "Chaos and Complexity in Richard Powers's *The Gold Bug Variations*." *Critique* 38.1 (1996): 38–51. Print.

———. "Home: A Conversation with Richard Powers and Tom LeClair." 2005. *Electronic Book Review*. 9 March 2008. Web. <http://www.electronicbookreview.com/thread/fictionspresent/corporate>.

Hickman, Jonathan, and J. M. Ringuet. *Transhuman*. Image, 2009. Print.

Holloway, Karla FC. *Private Bodies, Public Texts: Race, Gender, and a Cultural Bioethics*. Duke UP, 2011. Print.

Hoogheem, Andrew. "Secular Apocalypses: Darwinian Criticism and Atwoodian Floods." *Mosaic* 45.2 (2012): 55–71. Print.

Houser, Heather. "Wondrous Strange: Eco-Sickness, Emotion, and *The Echo Maker*." *American Literature* 84.2 (2012): 381–408. Print.

Hsu, Stephanie. "Ethnicity and the Biopolitics of Intersex in Jeffrey Eugenides's *Middlesex*." *MELUS* 36.3 (2011): 87–110. Print.

Hughes, Rowland. "The Ends of the Earth: Nature, Narrative, and Identity in Dystopian Film." *Critical Survey* 25.2 (2013): 22–39. Print.

Hume, Kathryn. "Moral Problematics in the Novels of Richard Powers." *Critique* 54 (2013): 1–17. Print.

Hungerford, Amy. *Postmodern Belief: American Literature and Religion since 1960*. Princeton UP, 2010. Print.

Huxley, Aldous. *Brave New World*. 1932. HarperCollins, 2006. Print.

Idema, Robert. "Toward a Minor Science Fiction: Literature, Science, and the Shock of the Biophysical." *Configurations* 23.1 (2015): 35–59. Print.

Interstellar. Dir. Christopher Nolan. Perf. Matthew McConaughey, Amy Adams. Paramount, Warner Bros., and Legendary, 2014.

In the Wake of the Flood. Dir. Ron Mann. Perf. Margaret Atwood. Sphinx, 2010.

Ishiguro, Kazuo. *Never Let Me Go*. Vintage, 2005. Print.

Jackson, Tom. "Richard Powers on Music, Twitter and His Novels." *Sandusky Register*. 9 Dec. 2014. Print.

Jameson, Fredric. "Then You Are Them." Rev. of *The Year of the Flood*, by Margaret Atwood. *London Review of Books* 31.17 (2009): 7–8. Print.

Johns, J. Adam. "Becoming Medusa: Octavia Butler's *Lilith's Brood* and Sociobiology." *Science Fiction Studies* 37.3 (2010): 382–400. Print.

Jones, Gwyneth. *Life*. Aqueduct, 2004. Print.

Jurecic, Ann. "The Art of Medicine: Life Writing in the Genomic Age." *Lancet* 383 (March 2014): 776–77. Print.

Kade, Stacey. *The Rules*. Hyperion, 2013. Print.

Kaufmann, Michael. "The Religious, the Secular, and Literary Studies." *New Literary History* 38.4 (2007): 607–27. Print.

Kevles, Daniel. *In the Name of Eugenics: Genetics and the Uses of Human Heredity*. Harvard UP, 1985. Print.

King, Stephen. *11/22/63*. Scribner, 2011. Print.

Kirby, David A., and Laura A. Gaither. "Genetic Coming of Age: Genomics, Enhancement, and Identity in Film." *New Literary History* 36.2 (2005): 263–82. Print.

Kress, Nancy. *Beggars and Choosers*. Tor, 1996. Print.

———. *Beggars in Spain*. William Morrow, 1993. Print.

———. *Beggars Ride*. Tor, 1994. Print.

Krokos, Dan. *False Memory*. Hyperion, 2012. Print.

Lake, Christina Bieber. *Prophets of the Posthuman: American Fiction, Biotechnology, and the Ethics of Personhood*. U of Notre Dame P, 2013. Print.

Lantos, John D. "A Better Life through Science?" *Hastings Center Report* 40.4 (2010): 22–25. Print.

Lee, Judith. "'We Are All Kin': Relatedness, Mortality, and the Paradox of Human Immortality." *Immortal Engines: Life Extension and Immortality in Science Fiction and Fantasy*. Ed. George

Slusser, Gary Westfahl, and Eric S. Rabkin. U of Georgia P, 1996. 170–82. Print.

Lee, Merton. "Why Jeffrey Eugenides's *Middlesex* Is So Inoffensive." *Critique* 51 (2010): 32–46. Print.

Le Guin, Ursula K. *Always Coming Home*. Harper & Row, 1985. Print.

———. *The Dispossessed*. 1974. Harper Voyager, 1994. Print.

———. *The Left Hand of Darkness*. Ace, 1969. Print.

———. "Nine Lives." 1969. *The Wesleyan Anthology of Science Fiction*. Ed. Arthur B. Evans et al. Wesleyan UP, 2010. 452–76. Print.

———. "The Ones Who Walk Away from Omelas." *The Wind's Twelve Quarters*. Harper & Row, 1975. 254–62. Print.

———. "Paradises Lost." *The Birthday of the World, and Other Stories*. Harper, 2002. 249–362. Print.

Lepore, Jill. "The History Lurking behind 'Orphan Black.'" *New Yorker*. 16 April 2015. Print.

Levin, Ira. *The Boys from Brazil*. Random House, 1976. Print.

Lewis, Sinclair. *Arrowsmith*. 1925. Signet, 1998. Print.

Logan. Dir. James Mangold. Perf. Hugh Jackman, Patrick Stewart, Dafne Keen. Donners', Kinberg Genre, and Marvel, 2017.

Mandel, Emily St. John. *Station Eleven*. Knopf, 2014. Print.

Marazano, Richard, and Jean-Michel Ponzio. *Genetiks*^{TM}. Vol 1. Boom, 2012. Print.

Markley, Robert. "Objectivity as Ideology: Boyle, Newton, and the Languages of Science." *Genre* 16.4 (1983): 355–72. Print.

Mawer, Simon. *Mendel's Dwarf*. 1998. Penguin, 1999. Print.

McEwan, Ian. *Saturday*. 2005. Vintage, 2006. Print.

McHugh, Susan. "Real Artificial: Tissue-Cultured Meat, Genetically Modified Farm Animals, and Fiction." *Configurations* 18.1–2 (2010): 181–97. Print.

McKibben, Bill. *Enough*. Times, 2003. Print.

McMann, Mindi. "British Black Box: A Return to Race and Science in Zadie Smith's *White Teeth*." *Modern Fiction Studies* 58.3 (2012): 616–36. Print.

Merchant, Carolyn. *The Death of Nature: Women, Ecology, and the Scientific Revolution*. Harper, 1980. Print.

Millington, Michael P. "Paneling Rage: The Loss of Deliberate Sequence." *Crossing Boundaries in Graphic Narrative*. Ed. Jake Jakaitis and James F. Wurtz. McFarland, 2012. 207–17. Print.

Mitchell, Robert. "Sacrifice, Individuation, and the Economies of Genomics." *Literature and Medicine* 26.1 (2007): 126–58. Print.

Mitchell, Robert, Helen Burgess, and Phillip Thurtle. *Biofutures: Owning Body Parts and Information*. DVD-ROM. U of Pennsylvania P, 2007.

Mitchell, W. J. T. *Cloning Terror: The War of Images, 9/11 to the Present*. U of Chicago P, 2011. Print.

Mitchison, Naomi. *Solution Three*. Warner, 1975. Print.

Moon. Dir. Duncan Jones. Perf. Sam Rockwell, Kevin Spacey. Liberty Films, 2009.

Moore, David S. *The Developing Genome: An Introduction to Behavioral Epigenetics*. Oxford UP, 2015. Print.

Morton, Timothy. *Hyperobjects: Philosophy and Ecology after the End of the World*. Minnesota UP, 2013. Print.

Mukherjee, Siddhartha. *The Gene: An Intimate History*. Scribner, 2016. Print.

Multiplicity. Dir. Harold Ramis. Perf. Michael Keaton, Andie MacDowell, Zack Duhame. Columbia, 1996.

Nanda, Aparajita. "Power, Politics, and Domestic Desire in Octavia Butler's *Lilith's Brood*." *Callaloo* 36.3 (2013): 773–88. Print.

Nelkin, Dorothy, and M. Susan Lindee. *The DNA Mystique: The Gene as a Cultural Icon*. U of Michigan P, 2004. Print.

Neuman, Birgit, and Ansgar Nünning. "Metanarration and Metafiction." *Living Handbook of Narratology*. Ed. Peter Hühn et al. Walter de Gruyter, 2009. Print.

Neuman, Justin. *Fiction beyond Secularism*. Northwestern UP, 2014. Print.

Newton, Maud. "Science, Chance, and Emotion with Real Cosima." *Longreads*. June 2015. Print.

Niccol, Andrew. *Gattaca* (Screenplay). Web. 19 Dec. 2016. <http://www.imsdb.com/scripts/Gattaca.html>.

Noble, David F. *The Religion of Technology: The Divinity of Man and the Spirit of Invention*. 1997. Penguin, 1999. Print.

Nowotny, Helga, and Guiseppe Testa. *Naked Genes: Reinventing the Human in the Molecular Age*. 2009. Trans. Mitch Cohen. MIT P, 2010. Print.

Obourn, Megan. "Octavia Butler's Disabled Futures." *Contemporary Literature* 54.1 (2013): 109–38. Print.

O'Connor, Flannery. *Mystery and Manners*. Farrar, Straus, and Giroux, 1957. Print.

O'Riordan, Kate. "Writing Biodigital Life: Personal Genomes and Digital Media." *Biography* 34.1 (2011): 119–31. Print.

Ostry, Elaine. "'Is He Still Human? Are You?': Young Adult Science Fiction in the Posthuman Age." *The Lion and the Unicorn* 28.2 (2004): 222–46. Print.

Parrington, John. *The Deeper Genome: Why There Is More to the Human Genome than Meets the Eye*. Oxford UP, 2015. Print.

Pinker, Stephen. "My Genome, My Self." *New York Times*. 7 Jan. 2009. Web. <http://www.nytimes.com/2009/01/11/magazine/11Genome-t.html>.

Powers, Richard. "The Book of Me." *GQ*. 30 Sept. 2008. Web. <http://www.gq.com/story/richard-powers-genome-sequence>.

———. "A Brief Take on Genetic Screening: Does Medicine Increase a Patient's Ability to Write Her Own Life?" *Believer* 4.2 (2006): 29–30, 46. Print.

———. *The Echo Maker*. Picador, 2006. Print.

———. *Generosity: An Enhancement*. Farrar, Straus, and Giroux, 2009. Print.

———. "Genie: New Byliner Fiction by National Book Award Winner Richard Powers." *PRWeb*. 9 Nov. 2012. Web. <http://www.prweb.com/releases/2012/11/prweb10108514.htm>.

———. *The Gold Bug Variations*. HarperCollins, 1991. Print.

———. *Orfeo: A Novel*. Norton, 2014. Print.

Query, Patrick R. "*Never Let Me Go* and the Horizons of the Novel." *Critique: Studies in Contemporary Fiction* 56.2 (2015): 155–72. Print.

"ReGenesis Behavioral Health." Web. 24 Aug. 2015. <http://regenesisonline.com>.

Regenesis Group Inc. Web. 24 Aug. 2015. <http://regenesisgroup.com/manifesto>.

ReGenesis Plastic Surgery and Skin Care Center. Web. 24 Aug. 2015. <http://regenesisplasticsurgery.com>.

Regenesis Skin Studio. Web. 24 Aug. 2015. <http://www.regenesisskinstudio.com>.

Rieder, John. "On Defining SF: Genre Theory, SF, and History." *Science Fiction Studies* 37.2 (2010): 191–209. Print.

Roof, Judith. *The Poetics of DNA*. Minnesota UP, 2007. Print.

Roth, Veronica. *Divergent*. HarperCollins, 2011. Print.

Rucka, Greg, Michael Lark, and Santi Arcas. *Lazarus*. Vol. 1. Image, 2013. Print.

Rush, Jennifer. *Altered*. Little, Brown, 2013. Print.

Russell, Mary Doria. *The Sparrow*. 1996. Ballantine, 1997. Print.

"Safe, Non-drug Pain Management." *ReGenesis Biomedical*. Web. 24 Aug. 2015. <http://www.regenesisbio.com/docs/514-0065-00%20D.pdf>.

Sargent, Pamela. *Cloned Lives*. Fawcett, 1976. Print.

Sawyer, Andy. "Kazuo Ishiguro's *Never Let Me Go* and 'Outsider Science Fiction.'" *Kazuo Ishiguro: New Critical Visions of the Novels*. Ed. Sebastian Groes and Barry Lewis. Palgrave Macmillan, 2011. 236–46. Print.

Schaefer, Heike. "The Pursuit of Happiness 2.0: Consumer Genetics, Social Media, and the Promise of Literary Innovation in Richard Powers' Novel *Generosity: An Enhancement*." *Ideas of Order: Narrative Patterns in the Novels of Richard Powers*. Ed. Antje Kley and Jan D. Kucharzewski. Winter, 2012. 263–84. Print.

Seaman, Myra J. "Becoming More than Human: Affective Posthumanisms, Past and Future." *Journal of Narrative Theory* 37.2 (2007): 246–75. Print.

Shaddox, Karl. "Genetic Considerations in Kazuo Ishiguro's *Never Let Me Go*." *Human Rights Quarterly* 35.2 (2013): 448–69. Print.

Shea, Elizabeth Parthenia. *How the Gene Got Its Groove: Figurative Language, Science, and the Rhetoric of the Real*. SUNY P, 2008. Print.

Sheldon, Rebekah. *The Child to Come: Life after the Human Catastrophe*. Minnesota UP, 2016. Print.

Shostak, Debra. "'Theory Uncompromised by Practicality': Hybridity in Jeffrey Eugenides's *Middlesex*." *Contemporary Literature* 49.3 (2008): 383–412. Print.

Slonczewski, Joan. "Octavia Butler's *Xenogenesis* Trilogy: A Biologist's Response." Paper presented at Science Fiction Research Association, Cleveland, Ohio. 30 June 2000. Web. <http://biology.kenyon.edu/slonc/books/butler1.html>.

Smith, Michael Marshall. *Spares*. 1996. Bantam, 1997. Print.

Smith, Rachel Greenwald. "Ecology beyond Ecology: Life after the Accident in Octavia Butler's *Xenogenesis* Trilogy." *Modern Fiction Studies* 55.3 (2009): 545–65. Print.

Smith, Zadie. *White Teeth*. Random House, 2000. Print.

Snyder, Katherine V. "'Time to Go': The Post-apocalyptic and the Post-traumatic in Margaret Atwood's *Oryx and Crake*." *Studies in the Novel* 43.4 (2011): 470–89. Print.

Springer, Katherine. "Hard Science Fiction in Film: Analyzing Duncan Jones's *Moon*." *Film Matters* (2012): 38–42. Print.

Squier, Susan Merrill. *Liminal Lives: Imagining the Human at the Frontiers of Biomedicine*. Duke UP, 2004. Print.

Stacey, Jackie. *The Cinematic Life of the Gene*. Duke UP, 2010. Print.

Stacy, Ivan. "Complicity in Dystopia: Failures of Witnessing in China Miéville's *The City and the City* and Kazuo Ishiguro's *Never Let Me Go*." *Partial Answers* 13.2 (2015): 225–50. Print.

Staes, Toon. "Dressing Up the Gene: Narrating Genetics in Richard Powers's *The Gold Bug Variations*." *LIT* 24.1 (2013): 48–64. Print.

Strecker, Trey. "Self-Organization and Selection in Richard Powers's *The Gold Bug Variations*." *Critique* 45.3 (2004): 227–45. Print.

Sun, Jian. "Interview: Fictional Collisions: Richard Powers on Hybrid Narrative and the Art of Stereoscopic Storytelling." *Critique* 54 (2013): 335–45. Print.

Taylor, Gordon Rattray. *The Biological Time Bomb*. New American Library World, 1968. Print.

Thurtle, Philip. *The Emergence of Genetic Rationality: Space, Time, and Information in American Biological Science, 1870–1920*. U of Washington P, 2007. Print.

Tillich, Paul. *Dynamics of Faith*. Harper, 1956. Print.

Tomasula, Steve. *VAS: An Opera in Flatland*. 2003. U of Chicago P, 2004. Print.

"The Trials." *ReGenesis*. Shaftsbury Films, Toronto. 21 Nov. 2004.

Trimm, Ryan S. "After the Century of Strangers: Hospitality and Crashing in Zadie Smith's *White Teeth*." *Contemporary Literature* 56.1 (2015): 145–72. Print.

Trushell, John M. "American Dreams of Mutants: The X-Men—'Pulp' Fiction, Science Fiction, and

Superheroes." *Journal of Popular Culture* 38.1 (2004): 149–68. Print.

Turner, Stephanie S. "Clone Mothers and Others: Uncanny Families." *SciFi in the Mind's Eye: Reading Science through Science Fiction*. Ed. Margret Grebowicz. Open Court, 2007. 101–13. Print.

Twelve Monkeys. Dir. Terry Gilliam. Perf. Bruce Willis, Madeleine Stowe, Brad Pitt. Universal Pictures, Atlas Entertainment, and Classico, 1995.

2001: A Space Odyssey. Dir. Stanley Kubrick. Perf. Keir Dullea, Gary Lockwood, William Sylvester. MGM, 1968.

van Dijck, José. *Imagenation: Popular Images of Genetics*. New York UP, 1998. Print.

Vaughan, Brian K., et al. *Y: The Last Man*. 2002–8. 5 vols. Deluxe ed. Vertigo, 2008–11. Print.

Venter, Craig. *Life at the Speed of Light: From the Double Helix to the Dawn of Digital Life*. Viking, 2013. Print.

Vint, Sherryl. "Commodifying Life: Posthumanism, Cloning and Gender in *Orphan Black*." *Science Fiction, Ethics, and the Human Condition*. Ed. Christine Cornea and Christian Baron. Springer, 2017. 95–113. Print.

———. "Why Have There Been No Great Women Scientists? Gender and Genetics in Gwyneth Jones's *Life*." *New York Review of Science Fiction* 22.2 (2009): 1, 8–13. Print.

Wald, Priscilla. "Future Perfect: Grammar, Genes, and Geography." *New Literary History* 31 (2000): 681–708. Print.

Waldby, Catherine, and Robert Mitchell. *Tissue Economies: Blood, Organs, and Cell Lines in Late Capitalism*. Duke UP, 2006. Print.

Wallace, Molly. "Reading *Xenogenesis* after Seattle." *Contemporary Literature* 50.1 (2009): 94–128. Print.

Ward, Graham. *True Religion*. Blackwell, 2003. Print.

Wasson, Sara. "'A Butcher's Shop Where the Meat Still Moved': Gothic Doubles, Organ Harvesting and Human Cloning." *Gothic Science Fiction, 1980–2010*. Ed. Sara Wasson and Emily Alder. Liverpool UP, 2011. 73–86. Print.

———. "Love in the Time of Cloning: Science Fictions of Transgressive Kinship." *Extrapolation* 45.2 (2004): 130–44. Print.

Watts, Jarica Linn. "'We Are Divided People, Aren't We?': The Politics of Multicultural Language and Dialect Crossing in Zadie Smith's *White Teeth*." *Textual Practice* 27.5 (2013): 851–74. Print.

Wells, H. G. *The Island of Doctor Moreau*. 1896. Penguin, 2005. Print.

———. *The War of the Worlds*. 1898. Vintage, 2017. Print.

Wexler, Alice. *Mapping Fate: A Memoir of Family, Risk, and Genetic Research*. U of California P, 1995. Print.

Wilhelm, Kate. *Where Late the Sweet Birds Sang*. Harper & Row, 1976. Print.

Wolfe, Gene. "The Fifth Head of Cerberus." 1972. *The Best of Gene Wolfe: A Definitive Retrospective of His Finest Short Fiction*. Tor, 2009. 31–77. Print.

Wolmark, Jenny. *Aliens and Others: Science Fiction, Feminism, and Postmodernism*. U of Iowa P, 1994. Print.

Wong, Cynthia F., and Grace Crummett. "A Conversation about Life and Art with Kazuo Ishiguro." 2006. *Conversations with Kazuo Ishiguro*. Ed. Brian W. Shaffer and Cynthia F. Wong. UP of Mississippi, 2008. 204–20. Print.

Žižek, Slavoj. *The Sublime Object of Ideology*. Verso, 1989. Print.

Index

absolutism, 44, 106, 134, 160, 210–11
academia, 8, 51, 125, 198
accident, 27, 81, 116, 119–26, 192. *See also* randomness
Adulthood Rites. See *Xenogenesis/Lilith's Brood* (Butler)
Africa, 209
Agamben, Giorgio, 233 n. 6
agency
    both individual and group, 14
    difficulty of exercising, 22, 221
    in genetic realism and genetic metafiction, 18
    limitation/denial of, 62, 74, 114, 124, 218
    optimism concerning, 100–101
    predisposed, 29, 101, 175–207, 209, 223
    in tension with accident/determinism, 27, 44, 115, 119–24, 211
    in tension with genetics and environment, 32–33, 36–38, 49, 71–72, 94, 143
agnosticism, 151. *See also* uncertainty
Alaimo, Stacy, 193
alien
    extraterrestrial, 23–24, 47, 62, 77–84, 93, 211–21
    ordinary life as, 187
    as other/unfamiliar, 51, 67, 69, 77, 83
    postapocalyptic world as, 175–77
allele, 37. *See also* nucleotide
Always Coming Home (Le Guin), 69

ambiguity
    as vice, 182
    as virtue, 20, 22, 79, 115, 121, 167
analogy, 9–14, 91, 210, 223
ancestry, 157, 188
animals. *See also* transgenic organism(s)
    as food sources, 153
    as ideals, 164
    inseparable from human beings, 23, 134, 137, 171, 181, 233 n. 6
    as reminders of creatureliness, 176–77
    as research subjects, 95, 120, 125–26, 143, 155, 230 n. 15
anthropocentrism, 137, 178
anthropology, 8, 67, 81, 108, 138, 234 n. 12
anthropomorphism, 123, 171, 181, 211
APOE allele, 37. *See also* disease: Alzheimer's
Aronofsky, Darren, 28
Arrival, 211–22
Arrowsmith (Lewis), 186–87
artificial insemination, 97–98
asexuality, 20, 143, 217, 231 n. 5
Asilomar Conference on Recombinant DNA, 26, 65, 95, 226 n. 2
Asimov, Isaac, 7
assimilation. *See also* colonialism
    between academic disciplines, 30, 39, 171, 203
    cultural, 78–82, 118–19, 128, 227 n. 10
astrological genetics/genomics, 16, 26, 37, 184

atheism, militant, 44–45, 51–52, 166, 171, 181, 222. *See also* New Atheists
Atwood, Margaret
    Blind Assassin, The, 157–59
    "Chicken Little Goes Too Far," 158
    Handmaid's Tale, The, 146–47, 151, 159
    "Happy Endings," 159
    MaddAddam trilogy, 28–29, 71, 134–35, 147–73, 175–77
    review of *The Echo Maker* (Powers), 177–78
    Scribbler Moon, 178, 233 n. 1
Augustine of Hippo, 200
authoritarianism, 54
autobiography, 43, 50, 106, 217–18, 233 n. 7, 234 n. 15
Avatar, 62, 143
awe. *See* wonder

Bacon, Francis, 52, 104
bacteria, 35, 190, 193–94, 203, 205
Barnes, Barry, 34, 37, 184
base pair, 33. *See* nucleotide
Battlestar Galactica, 62
Baudrillard, Jean, 73
Beggars trilogy (Kress), 230 n. 12
beliefs, 13, 45, 149, 180. *See also* faith
Bérubé, Michael, 26, 50, 106, 217–18
Bible
    "abomination," 98, 227 n. 6, 229 n. 4
    apocalypse in, 97, 162
    Daniel, 162
    Ecclesiastes, 114
    eco-friendly, 170
    fictional reference to, 28, 97–98, 101, 175, 210, 221–22
    Genesis, 52, 82–83, 140, 148, 232 n. 18
    Gospel of John, 25
    as "great code" (Frye), 59
    imitation of, 141
    as literal instruction manual, 40, 159
    "Lord's Prayer, The" 151
    as major text of world literature, 1, 59
    1 Samuel, 221, 235 n. 4
    Psalms, 52
    Revelation, 82, 162, 175, 201
    source of hope, 221
    as tool of authoritarianism, 43, 97
    use of names from, 59–60, 85, 228 n. 17
biblical archetype
    Adam figure, 25, 59, 83, 147–51, 160–72
    Christ figure, 69–70, 81–83, 142, 218, 221, 228 n. 16 (*see also* messianism)
    Eve figure, 25, 59, 83, 140, 150, 227 n. 12
    Mary figure, 59, 218
    Moses figure, 78, 168, 175
    Noah figure, 71, 148, 163–65
    Paul figure, 59, 72–76, 82, 228 n. 17
    Satan figure, 161
BiDil, 226 n. 5 (chap. 1)
big bang, 24
bildungsroman, 69, 73, 109
binary thinking
    about clones and humans, 68
    about epistemology and religion, 23–24, 36–37, 222
    about gender and sexuality, 116, 129
    about nature and nurture, 67
    about realism and metafiction, 199, 203–4
bioart, 183, 190
biocommerce, 232 n. 14
bioethics
    in classrooms, 31
    conflation of diverse issues within, 126

contributions of, 40, 65, 172
as interrogative rather than prescriptive, 229 n. 8
neglect of, 3, 19–20, 106
methodological purposes distinct from, 7, 21, 116
response to human reproductive cloning, 60
biohacking, 92, 183, 190, 203
bioinformatics, 35
biomedicine, 21, 32–33, 41, 116, 158, 210. *See also* medicine; personal medicine
biometrics, 27, 153. *See also* biopower
biopolitics, 13, 70, 216
biopower, 27, 31, 71
biotechnology companies, 6, 31, 43, 158, 225 n. 2 (chap. 1)
biotech slippage, 113, 126, 230 n. 15
bioweapon, 148, 150, 153, 162
birth, 32, 53–54, 56, 72, 108, 212
Blade Runner, 59, 62, 64, 221, 228 n. 22
Blair, Tony, 3, 11
blank slate motif, 61, 74, 89, 130. *See also* objectivism
Blind Assassin, The (Atwood), 157–59
blockbuster cinema, 10, 26, 60, 72, 111
blood, 22, 33, 103, 133, 205, 226 n. 5 (chap. 1)
Blueprint, 62
body. *See also* organ donation
vs. abstraction, 49–50
aging/limitation/death of, 76–77, 86
assimilation by informational network, 48, 168
context for gene expression, 12, 35
evaluation of, 56
as impure, 117
objectification of female, 141, 149–50, 154, 216 (*see also* objectification)
possible future forms of, 41
relationship to culture, 68
relationship to soul, 61–63, 80–84, 171, 177, 181
rewritten via bioengineering, 52
significance for identity, 5, 33
source of predisposition, 195
Boyle, Robert, 42
Boys from Brazil, The (Levin), 65, 70
brain
amygdala, 180
inferotemporal cortex, 180
as location of divine, 152–53
as mediator of sexual desire, 176
neocortex tissue, 155
relationship to mind, 124, 146, 181
as seat of soul, 104–5, 122, 135
as source of mind, 124, 146
Brave New World (Huxley), 10, 85, 92, 95–96
BRCA1 gene test, 40
Brooks, Peter, 124–25, 194–95, 214
Bush, George W., 45, 55
Butler, Octavia
"Evening and the Morning and the Night, The" 228 n. 13
Parable of the Sower, 80
Parable of the Talents, 80
Xenogenesis/Lilith's Brood, 26, 61, 77–84, 129, 142, 150

Camus, Albert, 202
Canada, 17, 103, 127, 169–70, 200, 229 n. 2
Canadian television, 17–23, 94–107, 170
capitalism, 39, 73–76, 95–97, 121, 155, 177–80
carbon-copy clone motif, 61–77, 85–89, 94, 121, 144, 147
Catholicism, 43, 45, 138–39
causation, 202, 209–13, 218–22, 234 n. 2
central dogma, 43. *See also* DNA

certainty. *See* uncertainty
chance. *See* accident
chemistry, 45, 91, 116, 166, 176, 185–92
Chicago, 186, 196, 201, 209
"Chicken Little Goes Too Far" (Atwood), 158
Childhood's End (Clarke), 168, 234 n. 3
children
    babies, 53–54, 72, 104, 122, 211, 220–22
    evaluation/commodification of, 53–56, 113–14, 154–55, 157
    experimentation on, 18, 72, 95, 101, 114
    relationship with parents, 94–99, 101–3, 106, 112–13, 142
chimera, 35, 119, 121
Christianity, 43, 171, 183, 226 n. 3, 228 n. 20, 233 n. 4. *See also* Catholicism; Protestantism
Church, George, 1–5, 11–12, 23, 183, 186, 225 n. 2 (chap. 1)
cinematography, 53
circular narrative, 71–74, 92, 125, 129, 209, 214. *See also* circular narrative; linear narrative; spiral narrative
civil religion, 53
Clarke, Arthur C., 7, 42, 168, 234 n. 3
classism, 13, 66, 87–89, 127–28, 157
Clayton, Jay, 41, 188, 197, 209, 225 n. 1 (chap. 1), 228 n. 18
climate change, 24, 40, 149, 154, 157–59, 220
Clinton, Bill, 3, 11
Cloned Lives (Sargent), 26–27, 61, 72–77, 100–101, 117, 125
cloning fiction
    bodies as empty shells in, 89, 111 (*see also* body)
    Carbon-Copy Clone Catastrophe, 27, 61–71, 73, 85–89, 92
    default maleness in, 70
    disease/early expiration in, 62, 86
    egg harvesting in, 97, 209
    incest in, 66, 74–75, 117, 133–34
    interchangeability in, 72–73, 115
    isolation in, 62, 67–69, 74–75, 109–10
    narcissism in, 74–75
    resurrection/reincarnation in, 69–70, 76–77, 79
    uniqueness/originality in, 72–73, 85, 107, 109–14
Cloning of Joanna May, The (Weldon), 71
closure. *See also* denouement
    overemphasis on, 21, 72
    possibilities for, 144
    resistance to traditional forms of, 167, 205–7, 214, 233 n. 8
Code 46, 44, 157
cognition, 177–81. *See also* brain
cognitive estrangement, 7
Colbert Report, The, 1, 186
Cold Spring Harbor Laboratory, 104–5, 229 n. 9
colonialism, 4, 8, 77–81, 118, 126–29
comics, 5, 7, 11–17, 28, 134–46, 231 n. 6
community, 66, 115, 156, 162, 166–69, 209–10
compositing, 29, 199
conception, 53–54, 73, 123, 221. *See also* birth
consciousness, 62, 122, 181, 202. *See also* soul
conservationism, 181
constrained constructivism, 49. *See also* Hayles, N. Katherine
constructedness, 86, 133, 200, 206, 209, 219
consumerism, 6, 87, 171, 176, 228 n. 20
Contact (Sagan), 162, 216
contingency, 144, 218. *See also* uncertainty

corporatism, 61, 85–89, 94–95, 101–3, 127, 153–62
cosmetic surgery, 5, 122, 231 n. 6
cosmology, 25, 193
Creation, The (Wilson), 171
creationism, 40, 45, 52, 233 n. 5. *See also* evolution; intelligent design
creative nonfiction, 29, 197, 201, 206–7, 209
Crick, Francis, 43–44
CRISPR, 4, 33, 225 n. 1 (intro.), 230 n. 16. *See also* gene editing
c-value enigma, 35

Darwin, Charles, 104, 170, 225 n. 5
Darwinian literary criticism, 234 n. 12. *See also* evocriticism
Dawkins, Richard, 44, 47–49, 123, 170–71, 175, 181
Dawn. See *Xenogenesis/Lilith's Brood* (Butler)
death drive, 141–42, 144
Delillo, Don, 172
denarration, 199
denouement, 206–7
Derrida, Jacques, 53
designer babies, 155, 157
determinism. *See* genetic determinism
Detroit, 116–17, 119, 123, 127–28, 133
deus ex machina, 93
Dickenson, Donna, 40, 226 n. 5 (chap. 1)
digitization, 41, 47–48, 130, 204, 226 n. 7
Dijck, José van, 26, 43–44, 48, 50
"Diploids—Die, Freak, The" (McLean), 59
disability, 14, 55, 76, 104–7, 111–12
disease
    acute myeloid leukemia (AML), 18–19, 22
    Alzheimer's, 37
    cancer (general), 39–40, 81, 106, 113, 121, 221
    Capgras syndrome, 180, 182 (*see also* brain)
    cerebral palsy, 121
    cystic fibrosis, 37
    depression, 31, 38, 74, 154
    as design fault, 152
    Down syndrome, 50, 106, 218
    vs. enhancement (in bioethics), 12
    follicular lymphoma, 103
    fragile X syndrome, 37
    Huntington's, 94, 228 n. 13
    influenza, 22
    muscular dystrophy, 37
    Parkinson's, 121
    progeria, 229 n. 1
    rhetoric of, 218
    single-gene disorder, 32, 37, 184
Dispossessed, The (Le Guin), 147
District 9, 157
Divergent series (Roth), 70
DNA, 1, 31–35, 50, 94, 115, 190–91
Do Androids Dream of Electric Sheep? (Dick), 59, 64. See also *Blade Runner*
Dog Stars, The (Heller), 191
Dolly the sheep, 95, 225 n. 7, 228 n. 22, 229 n. 10
doppelgänger, 65, 93, 136, 146
Dostoevsky, Fyodor, 114
double entendre, 139
dualism, 61, 84, 111, 150, 167. *See also* binary thinking
Dupré, John, 34, 37, 184
dystopia, 31, 56, 114, 147, 152

Echo Maker, The (Powers), 29, 177, 179–83, 194, 200, 233 n. 4
ecology
    collapse of, 71, 87, 168, 220
    effort to protect/balance, 6, 29, 143, 160, 176–80, 193–94
    water access, 182

Eden, Garden of, 25, 28, 41, 81, 139–40, 226 n. 6
editing (film/TV), 53, 101, 211–14, 219
education
    higher, 39 (*see also* academia)
    inequality in, 107–8
    K-12 science, 45, 187
    public, 17, 21–22, 48
edutainment (TV), 176, 188
11/22/63 (King), 69
Elmer Gantry (Lewis), 154, 162, 164
Elysium, 157
embryo, 5, 35–36, 52, 60, 96–97, 121
emotion. *See also* gender; subjectivity
    genetic influence on, 187, 201
    as strength, 168
    and subjectivity, 18–21, 41, 91, 147, 170
empathy, 64, 74, 79, 91–92, 186, 191
empiricism, 25, 29, 49, 187. *See also* objectivism
England, 87, 93, 117–18, 126, 231 n. 24
environment. *See also* ecology; epigenetics
    not "out there" but "in here," 191
    in tension with genetics and agency, 11–12, 37–38, 94, 100–101, 143, 201–2
environmentalism, 179–80, 193–94
environmental planning, 6
epigenetics, 35, 94–96, 130, 146–47, 155–56, 167
epistasis, 34
epistemology, 18–19, 45–51, 177, 182, 211, 217–19. *See also* faith; knowledge
essentialism, 42, 150, 178, 181
ethical, legal, and social implications (ELSI) of genetics, 40
ethnicity, 94, 117–18, 190. *See also* nationality; race
eugenics, 31–33, 53–57, 104–7, 124–28, 148, 217
Eugenics Record Office, 104–5, 229 n. 9
Eugenides, Jeffrey, 27, 94, 116–31, 133–34, 147, 230 n. 18
evangelicalism, 52, 233 n. 4. *See also* Christianity; Protestantism
"Evening and the Morning and the Night, The" (Butler), 228 n. 13
evocriticism, 234 n. 12. *See also* Darwinian literary criticism
evolution, 23–25, 45, 115–16, 168, 187, 209
existentialism, 74, 104, 108, 181
exogenesis, 24, 56
extinction, 3, 193
Eyes of Heisenberg, The (Herbert), 227 n. 6

faith, 42–46, 55–56, 152, 213. *See also* epistemology; intuition
fatalism, 40, 116, 178, 185, 189, 212. *See also* genetic determinism
Federal Bureau of Investigation (FBI), 183, 190
Fermat's principle, 234 n. 2
Feynman, Richard, 1
fiction, 134, 177, 198, 202. *See also* myth
"Fifth Head of Cerberus, The" (Wolfe), 227 n. 7
Fitter Families Contests, 104, 231 n. 6
5-alpha-reductase pseudohermaphroditism, 122
Foer, Jonathan Safran, 130
Fordism, 119, 127
foreknowledge, 37–39, 81, 163, 207, 211–14, 219–23. *See also* astrological genetics/genomics
fourth wall, breaking the, 210
FoxP2 gene, 152
Frankenstein; or, The Modern Prometheus (Shelley), 14, 68, 91–92, 95–98, 112–13, 152–53
Franklin, Benjamin, 141

Franklin, Rosalind, 44, 142
freedom, 80, 130, 168, 177, 198, 219–23. *See also* agency; genetic determinism
free indirect discourse, 64, 70, 198
Frye, Northrup, 59
funhouse mirror, 145, 159, 167
Future Library (Oslo, Norway), 178, 233 n. 1

Gadamer, Hans-Georg, 218
Gaddis, William, 234 n. 15
Gagnier, Regenia, 67
Gaia theory, 28, 78, 143
Galatea 2.2 (Powers), 197
Gattaca, 31–33, 50, 53–57
"gay gene," 36
Gelsinger, Jesse, 225 n. 9
gemmules, 225 n. 5
gender
    and comedy, 135–37
    and epistemology, 18–21, 218–19
    as fluid/flexible, 82–83, 116, 121–22, 129–30, 135
    and religion/myth, 15–16, 138–41, 150
    and technotranscendence, 9, 11
gene
    as composite object, 34
    deletion of, 151, 155
    demetaphorization of, 48
    personification of, 123
gene editing. *See also* gene therapy
    as computer programming, 126
    as divine pretension, 43, 70–71, 79–83, 96, 156–57, 217
    as path toward immortality, 151, 155
    religious motives of, 91
    side effects, 34
gene patenting, 101, 192
Generosity: An Enhancement (Powers), 29, 176–77, 182–92, 196–203, 206–7, 209

Genesis Code, The (Case), 62, 227 n. 6
gene therapy, 18–20, 39, 179. *See also* gene editing
genetic determinism. *See also* agency
    of astrological genetics, 16, 37
    in comics, 142, 146–47
    cultivation of circular narrative by, 120, 125
    as cultural determinism, 94, 109–10, 113–14, 126–27
    corporate forces driving, 100
    in film, 32–33
    generality of, 84
    resistance to, 70–75, 94, 123–24, 155, 189–90
    surrender to, 113–14
genetic discrimination, 31, 50, 53–57, 104–7, 124–28, 184–85. *See also* genetic determinism
genetic dismissivism, 16, 28, 39–40, 210, 220
genetic enhancement, 41, 78, 157–58, 199. *See also* gene editing; genetic modification
genetic fantasy, 9–17, 59–89, 107, 143–44, 159
genetic fiction
    astronomy/astronauts in, 56, 66–69, 72–73, 85–89, 193–94, 220
    cautionary elements of, 6–7, 17, 21–22, 152, 186, 232 n. 15
    comedy in, 98, 117, 133–36, 160–62, 232 n. 17
    complex scientist characters in, 103
    cryogenics in, 76, 172, 186
    evolutionary pattern of, 10–11, 209
    extinction of males in, 144
    feminism in, 97–98, 100, 103–4, 125–26, 137–46, 148–52
    hierarchy as problem in, 18, 77–79, 109, 217

genetic fiction (*continued*)
    history of, 85
    hope, cultivation of, 6–7
    impact on public attitudes, 32, 65, 85, 147
    incorporation of drama/theatre, 133, 135, 145
    influence on research funding/publication, 4, 45, 182, 187
    inspiration for biomedicine, 33
    irony in, 38
    language acquisition in, 91, 156
    mad scientist motif in, 91, 158, 186
    music in, 25, 116, 152, 183, 190–91, 195
    and nuclear physics/warfare, 43, 76, 78, 157, 229 n. 10
    origin myth in, 22, 57, 72, 123, 140, 147–48
    "playing God" in, 92
    post/transhuman characters in, 76, 82–83, 96, 148–54, 159, 166–68
    preconception/birth consciousness, 81, 116
    primates in, 138–39, 143, 148
    romance in, 23, 74–75, 117, 123, 135, 156, 198, 212
    sexism in, 15, 18–20, 71, 135, 138, 142
    suicide in, 73–74, 96, 101–2, 141–44, 182, 198–205
    technical specificity in, 103–4
genetic metafiction, 9–11, 22–25, 28–29, 133–73, 175–207
genetic modification, 2, 41, 106. See also gene editing
genetic realism, 9–11, 17–23, 27, 79, 84–85, 91–131
genetics
    do-it-yourself (DIY), 183, 190 (*see also* biohacking)
    and environment, 11, 61, 65, 79, 84
    and language, 2
    metaphorical significance of, 42, 147
    as pseudoreligion, 43, 106
genetic sublime, 11. *See also* technotranscendence
genetic testing
    affordability of, 93
    direct-to-consumer, 4, 31, 37, 43, 49, 197
    as divination, 93, 184
    exaggerated significance of, 55–56, 188, 202–3
    interpretation of, 32, 178, 195, 210–11
    as motive for early medical intervention, 40, 184
    optimism about, 2
    results as allegory for race/sex/gender/class/ability, 4
Genetiks™ (Marzano and Ponzio), 231 n. 6
genocide, 148, 153–54, 161–65
genome
    as "Book of Life," 50
    commercialization of, 49, 184, 197 (*see also* biotechnology companies; 23andMe)
    as digital information, 199
    as digital storage, 1
    as dynamic, 177, 184
    genome mosaicism, 36
    as a novel, 206–7
    as a site for divination, 93, 211
    size of, 31–35
    as source of soul, 43–44
genome time, 41, 188, 218. See also Clayton, Jay
genome-wide association study (GWAS), 35
genre, 7–8, 52, 197–99
geology, 52, 166
germline modification, 121. See also gene editing; genetic modification

gestation, 65, 74
Gibson, William, 234 n. 16
Gilman, Charlotte Perkins, 28
Girard, René, 89, 107, 109
global warming, 158. *See* climate change
"God gene," 36–37, 43
Godsend, 59
Gold Bug Variations, The (Powers), 85, 179, 197, 225 n. 7, 233 n. 2
golden ratio, 107. *See also* spiral narrative
Gould, Stephen Jay, 229 n. 8
GQ, 184
graphic novel/narrative. *See* comics
Green, Ronald M., 40, 172
GWAS. *See* genome-wide association study (GWAS)

Ham, Ken, 52. *See also* creationism
Hamer, Dean, 36–37, 42
Hamlet (Shakespeare), 104, 135, 138, 168, 173, 201
Handmaid's Tale, The (Atwood), 146–47, 151, 159
"happiness gene," 183, 187, 199
"Happy Endings" (Atwood), 159
Haraway, Donna, 43, 47, 67, 104, 125–26, 191–92
Hardin, Garrett, 154
Hardy, Thomas, 199, 233 n. 10
Hayles, N. Katherine, 26, 48–50
health and beauty industries, 5–6
Hebrew (language), 68, 218, 232 n. 18
Heidegger, Martin, 233 n. 6
Heinlein, Robert A., 7
heritability, 187
Herter, Cosima, 103, 229 n. 8
heterogeneity, 10, 28, 78, 80, 83, 92
heteronormativity, 19, 77, 136, 227 n. 8. *See also* homophobia
heterosexuality, 36, 227 n. 8
Hinduism, 126

history
  in academia, 44, 103, 182
  of bioethics, 65
  of cinema, 62
  erasure/repression of, 6, 41–42, 61, 94, 168
  of genetics, 226 n. 6, 229 n. 9
  of literature, 144
  of postsecular theory, 226 n. 10
  of science fiction, 9, 62
Hitler, Adolf, 70, 104, 125
HIV/AIDS, 21
homogeneity, 10, 28, 61, 73–76, 104, 189
homophobia, 61, 76, 82, 115, 136–37, 227 n. 8
Homo sapiens, 3, 35, 115, 141
homosexuality, 36, 227 n. 8. *See also* LGBTQ characters; queer identity
horror, 14, 197
House of the Scorpion, The (Farmer), 59, 155
Hughes, Langston, 114
Human Genome Project, 3–4, 10–11, 31–32, 44, 93–94, 225 n. 7
humanities, 5, 32, 39–44, 198
human subject protocol, 21
humility, 38, 96, 103, 151, 182, 192–94
hunch, 46, 187. *See also* faith; intuition
Hunger Games trilogy (Collins), 70
hybridity, 80–81, 91–92, 96–98, 116–20, 207, 223. *See also* purity fallacy
hyperbole, 3, 13, 36–37, 125, 185. *See also* genetic determinism
hyperobject, 149, 228 n. 21

ideology, 36–37, 42–50, 121, 125–26, 170–71, 210. *See also* absolutism; scientism
illness. *See* disease
Illumina, Inc., 32
Imago. *See Xenogenesis/Lilith's Brood* (Butler)
immeasurability, 49, 151, 153, 171, 181, 212. *See also* meaning; statistics

immigration, 89, 109, 115–20, 127–29, 185, 230–31 n. 18, 24
immunization, 40
industrialism, 9, 126–28, 147
inevitability, rhetoric of, 41, 57, 106, 113, 210. See also genetic determinism
information, 32, 48, 184–85, 211–12, 219, 222. See also digitization
injustice, 11, 66, 81, 87, 114, 128–29
In His Image (BeauSeigneur), 59, 81, 227 n. 6
instinct, 1, 98, 143, 151, 172, 181. See also animals; intuition
intelligent design, 23–24, 45, 144. See also creationism; evolution
interdisciplinarity, 5, 30, 39, 171, 195
interpretation, 29, 37, 121, 130, 197, 202–3. See also epistemology
interracial marriage, 117
intersexuality, 117, 122, 129
Interstellar, 234 n. 3
intertextuality, 92, 94–95, 101
intuition, 23, 45–46, 187, 195–96, 212–13. See also faith; hunch
Invasion of the Body Snatchers, 62–63, 227 n. 9
Invisible Man (Ellison), 8
in vitro fertilization (IVF), 55
Ishiguro, Kazuo, 27, 93–94, 107–16, 229–30 nn. 10–11
Islam, 25, 117, 119, 138, 231 n. 24
Islamism, militant, 15, 116, 138, 196
Island, The, 59, 64–65
Island of Doctor Moreau, The (Wells), 10, 91–92, 95–98, 102–3, 148–49
Israel, 139, 141, 148, 175, 229 n. 6. See also Hebrew (language); Judaism
IVF. See in vitro fertilization (IVF)

Jehovah's Witnesses, 117
Johns Hopkins University, 36, 231 n. 23
Jolie, Angelina, 40
Jones, Duncan, 27, 85, 87, 89, 107
Joshua Son of None (Friedman), 69–71
Joyce, James, 199, 233 n. 10
Judaism, 15, 51–52, 69, 138, 165, 221. See also Hebrew (language); Israel
junk DNA, 35
Jurassic World, 10
justice, 17, 89, 107–8, 135, 138, 143. See also injustice

Kalanithi, Paul, 221
Kennedy, John F., 69–70
Keystone XL pipeline, 157
King, Martin Luther, Jr., 12, 225 n. 6
King, Stephen, 69
knowledge, 23–25, 32, 42–46, 50–52, 192, 219. See also epistemology; faith
Korean (language), 85, 87
Kroeber, Alfred L., 67
Kurtz, Steve, 183
Kurzweil, Ray, 167

language
    ambiguity of, 169
    as dynamic metaphor, 49
    extrapolative capacity of, 171–72
    influence on thought, 234 n. 1
    limitations of, 164
    as source of relativism, 121
    subjectivity of, 2, 29, 39
    universal, 212
    written, 156, 212
Last Man, The (Shelley), 28, 135
Lazarus (Rucka, Lark, and Arcas), 231 n. 6
Left Hand of Darkness, The (Le Guin), 8, 116, 225 n. 2 (intro.)
Le Guin, Ursula K.
    Always Coming Home, 69
    Dispossessed, The, 147
    Left Hand of Darkness, The, 8, 116, 225 n. 2 (intro.)

"Nine Lives," 26, 61–71, 87–89, 92, 112, 231 n. 4
"Ones Who Walk Away from Omelas, The" 69
"Paradises Lost," 69
Lewis, Sinclair
   Arrowsmith, 186–87
   Elmer Gantry, 154, 162, 164
LGBTQ characters, 21, 27, 95, 136
Life (Jones), 115–16, 230 n. 17, 231 n. 4
Life of Pi (Martel), 8
Lilith's Brood/Xenogenesis (Butler), 26, 61, 77–84, 129, 142, 150
linear narrative, 92, 106–7, 129, 194–95, 209–22, 234 n. 2. See also circular narrative; linear narrative; spiral narrative
literalism, 40, 165–67, 232–33 n. 21. See also interpretation
Logan, 12
Long for This World (Byers), 229 n. 1
love
   as antithetical to use, 97–101, 108
   described by but exceeds biology, 95
   difference as crucial ingredient of, 92
   of enemies, 11
   as evolving addiction, 81–85
   grace as expression of, 68, 158
   of knowledge, 187
   materiality of, 166
   necessary expansiveness of, 23–25, 106
   as openness to other, 142
   outstripped by technology, 200
   queer form of, 27, 61, 66–69
   as self-sacrifice, 86–87
   transience of, 220–21
   as true fiction, 89
   unconditionality of, 94, 101

MaddAddam trilogy (Atwood), 28–29, 71, 134–35, 147–73, 175–77
magic, 14, 24, 42, 139, 141, 150

Maher, Bill, 52
Malcolm X, 12, 128, 225 n. 6
Man in the High Castle, The (Dick), 8
Manson, Graeme, 229 n. 2
Markley, Robert, 26, 42, 49–50
masculinity, performance of, 19, 82, 215–16
masterplot, 27, 61, 65, 71, 92
match cut, 209
materialism, 153, 162, 166, 168
materiality, 25, 27, 49, 151
McKibben, Bill, 157–58
meaning, 45–47, 184, 192–93, 200–202, 206–7, 222
measurement, 14, 37–39, 55, 166, 169, 187–90. See also empiricism; objectivism
medicine, 18–22, 40, 48, 106, 116, 210. See also genetics; personal medicine
Me Flows What You Call Time (Mitchell), 233 n. 1
Melville, Herman, 177
memory, 41, 62–64, 85, 171–72, 180–81, 212–14
Mendelian inheritance chart, 38
Mendel's Dwarf (Mawer), 229 n. 1
messianism, 47, 85, 138–39, 144, 183, 188. See also biblical archetype: Christ figure; scapegoat mechanism
metacinema, 211
metafiction. See genetic metafiction
metaphysical naturalism, 42, 202, 229 n. 8
metaphysical rhetoric, 3–4, 23, 43–45, 52
methodological naturalism, 42, 45, 202, 229 n. 8
Mexico, 17
microbiology, 24–25, 35, 163, 177, 183–84, 193–94
Middlesex (Eugenides), 27, 94, 116–31, 133–34, 147, 230 n. 18
military, 15, 102, 104, 211, 213, 215–17
Miller, Kenneth R., 45

Milton, John, 161, 164
mimesis, 9, 96, 109, 178, 199, 206. *See also* Girard, René
mise-en-abîme, 135, 144
mise-en-scène, 17, 214, 229 n. 2
mitochondrial DNA, 60
modest witness, 47, 186, 192, 194, 210, 227 n. 10. *See also* Haraway, Donna
Moon, 27, 85–89, 91
Morton, Timothy, 149, 228 n. 21
Muhammad, Elijah, 128
Multiplicity, 62–63, 65
mutation, 14, 81, 83, 115–16, 152, 177. *See also* chance; evolution
myth
    as delusion, 38, 66–67, 94, 110–11, 135, 150
    as foundation of culture, 2, 44, 52, 81–83, 155, 160
    living by, 57, 168, 195, 201, 213

narrative shape. *See* circular narrative; linear narrative; spiral narrative
National Book Award, 177
nationality, 27, 87, 94, 117–19, 213, 223. *See also* ethnicity
National Security Agency (NSA), 136
Nation of Islam, 116–18, 127
natural selection. *See* evolution
natureculture, 43, 67. *See also* Haraway, Donna
Nazism, 104, 124–26
neoliberalism, 39, 153
neutrality, 41, 51, 74, 218. *See also* epistemology; objectivism
Never Let Me Go (Ishiguro), 27, 93–94, 107–16, 229–30 nn. 10–11
New Atheists, 44, 47, 181. *See also* atheism, militant; secularism
Newton, Isaac, 42
Niccol, Andrew, 32, 55–56

9/11, 47, 55, 170
"Nine Lives" (Le Guin), 26, 61–71, 87–89, 92, 112, 231 n. 4
non-zero sum game, 213
novum, 8–9, 13, 76, 93, 142, 212. *See also* science fiction
NSA. *See* National Security Agency (NSA)
nucleotide, 25, 33–36, 191, 211. *See also* DNA
nurture, 60, 67, 109, 125, 130, 190. *See also* environment
Nye, Bill, 52

Oates, Joyce Carol, 234 n. 16
objectification, 82, 95, 107, 110–11, 141, 180. *See also* body
objectivism, 41–42, 45, 57, 89, 121, 125–26. *See also* relative objectivity; scientism
O'Connor, Flannery, 3
OncoMouse, 126
"Ones Who Walk Away from Omelas, The" (Le Guin), 69
Onion, The, 36. *See also* satire
Ontario Genomics, 21. *See also* biotechnology companies
Oprah, 188–89
Orfeo (Powers), 29, 183, 190–95, 203–7, 234 n. 16
organ donation, 68, 107–14, 155, 229 n. 11
Orphan Black, 27–28, 59, 93–108, 112–14, 133, 142
Oryx and Crake (Atwood). *See MaddAddam* trilogy (Atwood)
overpopulation, 28–29, 154, 163–64, 168, 175

Paley, William, 24, 144
palindrome, 160, 214
pandemic, 95, 135, 146, 148. *See also* disease; plague

Parable of the Sower (Butler), 80
Parable of the Talents (Butler), 80
paradigm shift, 25, 35, 45, 187, 230 n. 12. See also epistemology
paradise, 25, 188. See also Eden, Garden of
"Paradises Lost" (Le Guin), 69
patriarchy, 16–21, 54–55, 97–98, 120–21, 145–46
Pentecostalism, 219
perception, 9, 13, 26, 49–50, 188. See also interpretation; subjectivity
Percy, Walker, 162
personal medicine, 4, 93. See also genetic testing
personhood, 2, 7, 65, 156, 172, 226 n. 1. See also soul; subjectivity
PGD. See preimplantation genetic diagnosis (PGD)
phallic imagery, 94, 98, 101, 106, 136. See also psychoanalytic theory
pharmacogenomics, 2. See also personal medicine
plague, 135–38, 141–43, 148–50, 154–55, 164. See also disease; pandemic
Playboy, 66
pleiotropy, 37, 79
polygenic trait, 37, 79, 191
postapocalyptic setting, 71–72, 77–84, 134–68, 175, 232 n. 9
posthumanism, 41, 144, 188, 221
postsecular theory, 15, 45–57, 152, 210–11, 218–22, 226 n. 10. See also epistemology
Powers, Richard
   Echo Maker, The, 29, 177, 179–83, 194, 200, 233 n. 4
   Galatea 2.2, 197
   Generosity: An Enhancement, 29, 176–77, 182–92, 196–203, 206–7, 209
   genetic testing of, 32
   "Genie" (Powers), 23–25, 204
   Gold Bug Variations, The, 85, 179, 197, 225 n. 7, 233 n. 2
   Orfeo, 29, 183, 190–95, 203–7, 234 n. 16
Preformationism, 147, 225 n. 5
preimplantation genetic diagnosis (PGD), 52
previvors, 184
Priestley, Joseph, 52
Princess of Mars, A (Burroughs), 8
proof, 24, 47. See also epistemology; knowledge
proteomics, 4, 8, 33–36, 181, 229 n. 1
Protestantism, 43, 80, 169, 233 n. 4
pseudoscience, 139, 142–43, 156, 159, 190, 231 n. 4
psychoanalytic theory, 75, 215, 229 n. 5
Pulp Fiction, 141
purity fallacy, 15, 94, 116–19, 165. See also hybridity

queer identity, 36, 75–78, 82, 99–100, 227 n. 8

race, 13, 51, 78–80, 87, 124, 127–29. See also ethnicity
racism, 78–80, 87, 116–19, 126–28. See also colonialism
randomness, 24, 139. See also accident
rationalism, 152–53. See also certainty; uncertainty
realism. See genetic realism
reductionism, 32, 100, 130, 177, 181, 191. See also empiricism; rationalism
Regenesis (Cherryh), 225 n. 10
ReGenesis (TV series), 17–22, 97
Reimer, David, 231 n. 23
relative objectivity, 29, 45, 167, 182, 226 n. 8. See also epistemology; subjectivity

Index

relativism, 182
religion
- definition of, 52
- as dogmatism, 42–43, 46, 180, 207, 218
- and ecology, 161
- fundamentalism, 15–16, 51–52, 151, 165, 170–71, 232–33 n. 21
- metaphysics, 23–24, 29, 76, 229 n. 8
- and science, 46, 166, 171
- sexism in, 139
- as special effect, 53
- as ultimate concern, 57 (see also Tillich, Paul)

Religulous, 52
repetition, 57, 83, 85–86, 112, 195, 209–10. *See also* circular narrative
replication, 107, 115–16, 147, 177, 191, 230 n. 14. *See also* repetition
reproduction, 77, 101, 140, 148, 211, 215–18. *See also* birth; children
reverence, 68, 179, 206. *See also* humility; wonder
revision, 83, 185–86, 202, 206, 215. *See also* evolution
RNA interference (RNAi), 19
Robinson, Kim Stanley, 177, 220
Romanticism, 175
Rushdie, Salman, 234 n. 16
Russ, Joanna, 28

Sacks, Oliver, 180, 186
sacrifice, 85–89, 103, 108, 112
Sade, Marquis de, 141
Sagan, Carl, 162, 216
samsara, 119. *See also* Hinduism
Sapir-Whorfian hypothesis, 234 n. 1
satire, 36, 38, 71, 93, 157–61, 175–76
Saturday (McEwan), 94
scapegoat mechanism, 78, 85–89, 107–9, 113–14, 185, 196. *See also* Girard, René
science fiction
- AI/robots in, 71, 85, 115
- alternate history, 27, 93, 104, 107
- cyberpunk, 9, 53
- feminist, 26, 60–61, 70, 92
- first-contact, 23, 162, 211, 234 n. 3
- Golden Age, 9
- New Wave, 9, 70
- pulp, 9
- young adult, 226 n. 5 (chap. 2)

scientism, 42, 70, 116, 190–93, 203, 217. *See also* metaphysical naturalism; objectivism
SCNT. *See* somatic cell nuclear transfer (SCNT)
Scott, Ridley, 62
secularism, 47, 50–53, 76, 166. *See also* postsecular theory; secularization theory
secularization theory, 55, 234 n. 12
Selfish Gene, The (Dawkins), 44, 48, 123
sentimentality, 68, 99, 148, 161, 181–82, 191
sequence, 57, 195, 212, 214–15, 219–22
serpent wisdom, 28, 149–50. See also *MaddAddam* trilogy (Atwood)
Shelley, Mary
- *Frankenstein; or, The Modern Prometheus*, 14, 68, 91–92, 95–98, 112–13, 152–53
- *Last Man, The*, 28, 135

"Ship Fever" (Barrett), 229 n. 1
simultaneity, 57, 195, 209, 212, 214, 219–22
single nucleotide polymorphism (SNP), 34–35, 188
6th Day, The, 59
slavery, 78, 81, 86, 107–8, 150, 192. *See also* colonialism
slipstream, 8, 27. *See also* science fiction
Slonczewski, Joan, 228 n. 14
Smith, Zadie, 27, 94, 116–27, 146, 152
SNP. *See* single nucleotide polymorphism (SNP)

sociobiology, 84, 143, 228 n. 21
Solution Three (Mitchison), 59, 227 n. 6
somatic cell nuclear transfer (SCNT), 60
soul
    absence of, 71
    communal nature of, 78
    dignity of, 21
    discovery through loss, 205
    im/mortality of, 27
Soylent green, 154
Spares (Smith), 64, 155, 227 n. 6
Sparrow, The (Russell), 234 n. 3
spectacle, 10, 51, 64, 108, 119, 189–90
speculative fiction, 7–8. *See also* science fiction
spiral narrative, 5, 57, 106–7, 124–26, 129, 214–15. *See also* circular narrative; linear narrative; spiral narrative
spirituality, 139, 153, 155, 158, 181. *See also* meaning; soul
Splice, 14
Squier, Susan, 4–5, 32, 39, 43
Station Eleven (Mandel), 164–65
statistics, 32–37, 53, 171, 181, 211, 222. *See also* empiricism; relative objectivity
status quo bias, 172
stem cell research, 2, 60, 96, 103, 106, 121
Stepford Wives, The (Levin), 71, 135, 145
"Story of Your Life" (Chiang), 211–12, 222, 234 n. 2
subjectivity, 18–21, 62, 182, 202, 217–22. *See also* agency; epistemology
suburbia, 62, 95, 128, 153, 175
sufficient warrant, 29. *See also* epistemology
surveillance culture, 27, 80, 153. *See also* biometrics; biopower
suspension of disbelief, 9, 134
Suvin, Darko, 7
symbiology, 67

symbolic behavior, 156. *See also* language
synecdoche, 44, 111, 149
synergy, 83, 228 n. 19
synthetic biology, 2, 44–45, 176, 206, 210
systemic injustice, 128. *See also* injustice

Taoism, 69, 226 n. 3
Taylor, Gordon Rattray, 231 n. 4
technoculture, 8, 43, 59, 128
technotranscendence, 9, 42, 52, 75–77, 106–7, 210
television, 1, 17–25, 95–107, 182, 188–90
Teller, Edward, 188
terrorism
    bioterrorism, 95, 153, 183–84, 191
    ecoterrorism, 163
    fear of, 190–91
theology
    apophatic, 150–52
    cataphatic, 150–51
    and comedy, 2
    and gender, 139–41
    limitations of, 164
    and science, 40–43, 52, 80–84, 152, 165–69, 218
Thoreau, Henry David, 158
thought experiment, 7, 24, 169. *See also* science fiction
Three-Body Problem, The (Cixin), 234 n. 3
356.org, 157. *See also* McKibben, Bill
thriller, 20, 60–64, 76, 81–84, 92–93, 99. *See also* blockbuster cinema; spectacle
Tillich, Paul, 53
Toronto, 95, 229 n. 2
tragedy, 67, 168, 205, 220, 232 n. 12
tragicomedy, 135–37, 145, 168, 178
transcendence. *See* technotranscendence
trans-corporeality, 193
transgenic organism(s), 2–3, 14, 91, 96–97, 102, 158. *See also* animals; hybridity

Transhuman (Hickman and Ringuet), 231 n. 6
transience, 114, 172, 178, 205, 210, 221
Tropic of Orange (Yamashita), 8
true fiction, 61, 89. *See also* myth
Truman Show, The, 86
Twelve Monkeys, 47
23andMe, 31, 225 n. 2 (chap. 1). *See also* genetic testing
2001: A Space Odyssey, 85, 215, 228 n. 22, 234 n. 3

uncertainty
    attempt to control/eliminate, 120, 126, 179, 184–85
    comfort with, 21, 115–16, 158, 195
    as dynamic, 80–81
    vs. performed certainty, 106, 116, 125, 182, 185, 222
    as space of risk, 215–19
    value of, 47, 49–52
United Kingdom, The, 87, 168–69
United States, The, 17, 46, 51, 65, 157, 169–70
unreliable narration, 185
utilitarianism, 50, 73, 97, 100
utopia, 66, 76–77, 107, 114, 147, 159

VAS: An Opera in Flatland (Tomasula), 231 n. 6
Venter, Craig, 4, 44–48, 188, 226 n. 7
Viagra, 141, 161
Villeneuve, Denis, 211–22
violence, 54–55, 82, 98–101, 149, 168–70
virus. *See* disease; pandemic
vitalism, 45–46, 226 n. 7

Wald, Priscilla, 26, 41, 44, 50
War and Peace (Tolstoy), 144
Ward, Graham, 53
War of the Worlds, The (Wells), 234 n. 3
Watergate, 65

Watson, James, 43–44, 56, 71
Wells, H. G.
    Island of Doctor Moreau, The, 10, 91–92, 95–98, 102–3, 148–49
    War of the Worlds, The, 234 n. 3
Where Late the Sweet Birds Sang (Wilhelm), 59, 65, 71–72, 232 n. 19
White Plague, The (Herbert), 232 n. 11
White Teeth (Smith), 27, 94, 116–27, 146, 152
whole exome sequencing, 225 n. 2 (chap. 1)
whole genome sequencing, 32, 38, 184
Wilbur, Richard, 25
Wilson, E. O., 171, 228 n. 21
Wittgenstein, Ludwig, 25
Wöhler, Friedrich, 45
Womb, 62
wonder. *See also* awe; reverence
    at the earthly/material/ordinary, 177–79, 187, 194
    at microscopic life, 20–21, 97
    at new biotech, 1, 9, 223
    at telescopic vision, 193
Wright, Frank Lloyd, 118

Xenogenesis/Lilith's Brood (Butler), 26, 61, 77–84, 129, 142, 150
X-Files, The, 59, 65
X-Men, 11–17, 21, 23, 84, 225 n. 6
X-Men 2, 12
X-Men Regenesis, 12–16

Year of the Flood, The (Atwood). *See MaddAddam* trilogy (Atwood)
Y: The Last Man (Vaughan, Guerra), 28, 134–46, 149, 159, 217–18

Zeno's paradox, 213
Zero K (Delillo), 172
Žižek, Slavoj, 47
zoomorphism, 181